The World of
HORTICULTURE

THE AUTHOR

Dr. Shalu Vyas is a distinguished author and academician, possessing an impressive background in horticulture and agriculture. She completed her Ph.D. at Chaudhary Charan Singh Haryana Agricultural University (CCSHAU) in Hisar, Haryana, specializing from department of Horticulture. Her academic journey has been marked by excellence and recognition, with several notable achievements in research, teaching, and agricultural extension activities. With a total of 10 years of experience, Dr. Shalu Vyas has demonstrated her commitment to advancing the field of horticulture through her research contributions. Her research papers have been published in prestigious national and international journals, showcasing her expertise and valuable insights in various aspects of horticulture and agricultural sciences. In addition to her research work, Dr. Shalu Vyas has actively engaged in agricultural extension activities, through her involvement in extension activities, she has worked closely with farmers and rural communities, disseminating knowledge about modern agricultural practices, and promoting sustainable farming methods. Her efforts in this domain reflect her dedication to bridging the gap between academic research and real-world agricultural challenges. Dr. Shalu Vyas has been honored with the Young Scientist Award in an international conference in collaboration with NITI Aayog, recognizing her exemplary contributions to the field of horticulture and research. Such prestigious recognition further reinforces her standing as an influential figure in her field. Even during her B.Sc. program, Dr. Shalu Vyas received the Puran Anand Adhlkha Fellowship, a testament to her exceptional academic performance and dedication to her studies.Her contributions and achievements continue to inspire students, researchers, and fellow academicians alike.

The World of
HORTICULTURE

Dr. Shalu Vyas
(Assistant Professor)
Department of Horticulture,
Sant Baba Bhag Singh University,
JalandharPunjab-144030

2024
Daya Publishing House®
A Division of
Astral International Pvt. Ltd.
New Delhi – 110 002

ISBN: 978-93-5461-999-1

Published by : **Daya Publishing House®**
A Division of
Astral International Pvt. Ltd.
– ISO 9001:2015 Certified Company –
4736/23, Ansari Road, Darya Ganj
New Delhi-110 002
Ph. 011-43549197, 23278134
e-mail: info@astralint.com
Website: www.astralint.com

Printed at : **Replika Press Pvt Ltd**

Preface

Welcome to **"The World of Horticulture"** This book is a journey into the captivating world of horticulture, a field that merges science, art, and nature's wonders to cultivate plants and shape landscapes. Horticulture is a rich tapestry of knowledge that spans across disciplines such as biology, agronomy, botany, environmental science, and design. It touches every aspect of our lives, from the food we eat to the beauty that surrounds us. In this book, we embark on a systematic exploration of horticulture, unraveling its principles, practices, and profound impact on the world. Our journey begins with the foundational concepts of plant anatomy, physiology, and genetics. We delve into the marvels of plant growth and development, uncovering the intricate mechanisms that govern everything from germination to fruiting. As we traverse the chapters, you will encounter the diverse techniques that horticulturists employ to propagate, nurture, and enhance plant species. From traditional methods to cutting-edge technologies, each approach has a role to play in the art of cultivation. Landscapes are living canvases, shaped by both nature's hand and human creativity. Within these pages, we explore the art of designing gardens and green spaces, considering aesthetics, functionality, and sustainability. We also delve into the nuances of soil science, irrigation, pest management, and the delicate balance between maintaining plant health and protecting the environment.

Embark on this journey with curiosity and an open heart, and may the knowledge within these chapters inspire you to sow seeds of growth, beauty, and sustainability in your own corner of the world. Your valuable suggestions will go a long way to achieve this end, such suggestions will thus be warmly appreciated and heartly welcomed.

Happy gardening!

Dr. Shalu Vyas

Contents

General Horticulture

1

Introduction to Horticulture

Horticulture is a fascinating and intricate field that revolves around the cultivation, management, and utilization of plants for various purposes. Derived from the Latin words *"hortus"* (garden) and *"cultura"* (cultivation), horticulture encompasses a wide array of activities ranging from growing fruits, vegetables, and ornamental plants to landscape design and plant conservation. The world of horticulture is a captivating tapestry of plant cultivation, scientific discovery, and artistic expression. It encompasses a rich array of practices, from nurturing orchards laden with succulent fruits to designing intricate landscapes that awaken the senses. Horticulture's significance resonates across cultures and continents, touching upon essential aspects of human life.

At its core, horticulture is about harnessing the marvels of plant life. It involves understanding the intricate dance between soil, water, sunlight, and nutrients that sustains the growth of diverse crops. From the meticulously tended rows of a vegetable garden to the carefully crafted patterns of a botanical garden, horticulture is a symphony of biological interactions that yields sustenance, beauty, and scientific insight.

In recent times, the world of horticulture has witnessed remarkable innovation, with technological advances such as precision agriculture, vertical farming, and hydroponics revolutionizing how plants are cultivated. These innovations, coupled with a renewed focus on sustainability and conservation, showcase horticulture's capacity to address pressing global challenges such as food security and environmental degradation. As the world continues to evolve, horticulture remains a steadfast companion, offering a blend of tradition and modernity that enhances our connection with nature and the very essence of life.

Significance of Horticulture

Horticulture holds immense significance in various aspects of human life, ranging from nutrition and economy to environmental conservation and cultural aesthetics. Its multifaceted contributions make it a vital field that addresses critical global challenges. In this section, we delve into the profound significance of horticulture and its impact on different spheres of society.

- ✯ **Food Security and Nutrition:** Horticulture plays a pivotal role in ensuring food security and providing a diverse range of nutritious foods. Fruits, vegetables, and other horticultural crops are rich sources of vitamins, minerals, fiber, and antioxidants that are essential for maintaining good health. The cultivation of these crops helps diversify diets and reduces the risk of malnutrition, contributing to the overall well-being of individuals and communities.
- ✯ **Economic Development:** The horticultural industry has a substantial economic impact, creating employment opportunities and driving agribusiness. From small-scale local markets to large commercial enterprises, horticulture provides income for farmers, laborers, distributors, retailers, and researchers. The export of horticultural products contributes significantly to the economies of many countries, strengthening their trade and fostering rural development.
- ✯ **Environmental Conservation:** Horticulture has a role in environmental conservation by promoting sustainable practices. Well-designed landscapes and urban green spaces aid in carbon sequestration, air purification, and temperature regulation. Moreover, sustainable horticultural practices, such as organic farming and integrated pest management, reduce the use of harmful chemicals and minimize soil degradation, contributing to long-term environmental health.
- ✯ **Cultural and Aesthetic Value:** The beauty and aesthetics of horticulture enrich our surroundings and enhance the quality of life. Gardens, parks, botanical collections, and urban landscapes provide spaces for relaxation, recreation, and cultural activities. Horticulture's influence on art, literature, and cultural traditions demonstrates its deep-rooted connection to human creativity and expression.
- ✯ **Biodiversity Conservation:** Horticulture contributes to the conservation of plant biodiversity by preserving and propagating rare, endangered, and indigenous plant species. Horticulturists often work to maintain genetic diversity through seed banks, botanical gardens, and specialized

cultivation methods. This helps safeguard plants that may have ecological, cultural, or medicinal importance.

- ★ **Research and Innovation:** Horticulture serves as a fertile ground for scientific research and innovation. Advances in plant breeding, genetics, and biotechnology have led to the development of improved crop varieties that are more resistant to pests, diseases, and environmental stressors. These innovations address the challenges posed by a changing climate and limited resources.
- ★ **Health and Well-being:** Engagement with horticulture has been linked to positive health outcomes. Gardening and interacting with plants have been shown to reduce stress, improve mental well-being, and promote physical activity. Therapeutic horticulture programs use plant-based activities to aid in the rehabilitation of individuals with physical, mental, or emotional challenges.
- ★ **Education and Knowledge Sharing:** Horticulture serves as an educational platform, disseminating knowledge about plant biology, environmental stewardship, and sustainable practices. Schools, universities, and extension programs offer opportunities for students and communities to learn about cultivation techniques, plant biology, and the importance of responsible land use.

In conclusion, the significance of horticulture is multifaceted and far-reaching. From addressing global food challenges to contributing to economic development, environmental conservation, and human well-being, horticulture's impact is woven deeply into the fabric of society. Its ability to intersect with science, culture, and economy positions horticulture as a cornerstone of sustainable development and a key player in shaping a healthier and more harmonious world.

Historical Evolution

The roots of horticulture trace back thousands of years to ancient civilizations such as the Egyptians, Greeks, and Romans, who cultivated gardens for sustenance and aesthetic pleasure. Over time, horticulture advanced as a scientific discipline, with breakthroughs in plant breeding, propagation, and pest management. The 19th and 20th centuries witnessed significant strides in horticultural research and technology, leading to the development of new plant varieties, improved cultivation techniques, and better understanding of plant physiology.

Branches of Horticulture

Horticulture encompasses several specialized branches, each focused on specific aspects of plant cultivation and utilization. Some of the major branches include:

- **Pomology:** This branch deals with the cultivation of fruit-bearing trees and shrubs. Pomologists work on improving fruit quality, developing new cultivars, and managing orchard ecosystems.
- **Olericulture:** Olericulture focuses on the production of vegetables. Olericulturists study various aspects of vegetable cultivation, including crop rotation, soil management, and pest control.
- **Floriculture:** Floriculture involves the cultivation of ornamental and flowering plants for decorative purposes. This branch contributes to the beauty of landscapes, gardens, and indoor spaces.
- **Landscape Horticulture:** Landscape horticulturists design, install, and maintain outdoor spaces, combining elements of art and science to create visually appealing and functional environments.
- **Medicinal and Aromatic Plants:** This branch centers on the cultivation of plants used for medicinal, culinary, and aromatic purposes. It involves studying the chemical compounds responsible for their beneficial properties.
- **Postharvest Physiology:** Postharvest physiologists focus on extending the shelf life and quality of horticultural products after they have been harvested. This involves storage, transportation, and handling techniques.
- **Urban Horticulture:** Urban horticulture addresses the challenges of growing plants in urban environments, aiming to improve air quality, enhance aesthetics, and promote green spaces in cities.

Interdisciplinary Nature

Horticulture's dynamic and multidisciplinary nature lies at the heart of its success in addressing complex challenges related to plant cultivation, sustainability, and human well-being. Rooted in the intimate understanding of plants, horticulture draws upon a diverse array of scientific fields to optimize cultivation practices, develop new varieties, and innovate sustainable techniques. Botany forms the foundational science, providing insights into plant physiology, growth patterns, and genetics. These botanical principles, in turn, intersect with various disciplines, creating a collaborative ecosystem of knowledge.

Genetics and biotechnology contribute significantly to horticulture's interdisciplinary landscape. Plant breeding, a cornerstone of horticultural advancement, combines genetics with agriculture to develop improved crop varieties with desirable traits such as disease resistance, yield, and nutritional content. Genetic engineering further widens the scope, enabling the modification of plants to withstand environmental stressors or produce specific compounds, such as vitamins or pharmaceuticals. Environmental sciences, including soil science and meteorology, play a vital role in sustainable horticulture. Understanding soil composition and nutrient availability guides optimal cultivation techniques, while meteorology helps predict weather patterns and mitigate the impact of climatic fluctuations. Integrated pest management, another interdisciplinary approach, merges entomology and ecology to manage pests and diseases without relying solely on chemical treatments.

Horticulture thrives at the crossroads of various scientific fields, demonstrating its interdisciplinary essence. By integrating botanical knowledge, genetics, biotechnology, and environmental sciences, horticulturists drive innovation, foster sustainability, and contribute to the global effort of providing nutritious food, preserving biodiversity, and creating aesthetically pleasing environments.

Horticulture's Interdisciplinary Nature

Horticulture's dynamic and multidisciplinary nature lies at the heart of its success in addressing complex challenges related to plant cultivation, sustainability, and human well-being. Rooted in the intimate understanding of plants, horticulture draws upon a diverse array of scientific fields to optimize cultivation practices, develop new varieties, and innovate sustainable techniques. Botany forms the foundational science, providing insights into plant physiology, growth patterns, and genetics. These botanical principles, in turn, intersect with various disciplines, creating a collaborative ecosystem of knowledge.

Genetics and biotechnology contribute significantly to horticulture's interdisciplinary landscape. Plant breeding, a cornerstone of horticultural advancement, combines genetics with agriculture to develop improved crop varieties with desirable traits such as disease resistance, yield, and nutritional content. Genetic engineering further widens the scope, enabling the modification of plants to withstand environmental stressors or produce specific compounds, such as vitamins or pharmaceuticals.

Environmental sciences, including soil science and meteorology, play a vital role in sustainable horticulture. Understanding soil composition and nutrient availability guides optimal cultivation techniques, while meteorology helps predict weather patterns and mitigate the impact of climatic fluctuations. Integrated pest management, another interdisciplinary approach, merges entomology and ecology to manage pests and diseases without relying solely on chemical treatments.

Horticulture thrives at the crossroads of various scientific fields, demonstrating its interdisciplinary essence. By integrating botanical knowledge, genetics, biotechnology, and environmental sciences, horticulturists drive innovation, foster sustainability, and contribute to the global effort of providing nutritious food, preserving biodiversity, and creating aesthetically pleasing environments.

Future Trends

The future of horticulture holds exciting possibilities, including advancements in sustainable practices, utilization of vertical farming, and the integration of digital technologies for precision cultivation. As the global population continues to grow, horticulture will play a critical role in ensuring food security, environmental conservation, and the aesthetic enhancement of our surroundings.

In conclusion, horticulture is a dynamic and essential field that encompasses a wide range of activities aimed at cultivating plants for various purposes. Its historical evolution, diverse branches, and interdisciplinary nature make it a captivating subject with far-reaching impacts on human life and the environment.

2

Basics of Plant Biology

Plants are complex organisms that play a vital role in ecosystems, food production, and the environment. This chapter provides an overview of the basics of plant biology, covering topics such as plant structure, growth, reproduction, and the physiological processes that govern plant life.

Plant Structure: Exploring the Anatomy of Plants

Plants possess a fascinating and intricate structure that reflects their adaptation to various environments and roles within ecosystems. Understanding the different parts of a plant and their functions is essential for horticulturalists, botanists, and anyone engaged in plant-related activities.

- **Roots:** The root system anchors the plant in the soil and plays a crucial role in water and nutrient absorption. Roots are composed of primary and lateral roots, with root hairs increasing surface area for efficient nutrient uptake.
- **Stems:** Stems provide structural support, transport water, minerals, and nutrients throughout the plant, and serve as the sites for photosynthesis in some species. They contain vascular tissues—xylem for water transport and phloem for nutrient transport.
- **Leaves:** Leaves are the primary sites of photosynthesis. They contain chloroplasts, where chlorophyll captures sunlight for energy conversion. Leaf anatomy includes the upper epidermis, mesophyll layer, and lower epidermis, along with stomata that regulate gas exchange.

- ★ **Flowers:** Flowers are reproductive structures that facilitate pollination and seed production. They consist of various parts, including sepals, petals, stamens (male reproductive organs), and carpels (female reproductive organs).
- ★ **Fruits:** Fruits develop from fertilized ovaries of flowers and protect seeds while aiding in their dispersal. They exhibit diverse forms, from fleshy fruits like apples to dry fruits like nuts.
- ★ **Seeds:** Seeds contain the embryonic plant, stored nutrients, and a protective seed coat. Germination, triggered by water, warmth, and sometimes light, initiates growth.

Plant Growth and Development

The journey of a plant from a tiny seed to a mature individual is a complex process governed by intricate cellular mechanisms and external factors. Plant growth and development encompass a series of dynamic changes that shape the plant's structure and function throughout its life cycle.

- ★ **Cell Division and Growth:** Plant growth starts with cell division, occurring in meristematic tissues found at the tips of roots and shoots. These cells divide and differentiate, leading to elongation, which drives the plant's increase in size.
- ★ **Differentiation and Specialization:** As cells divide, they become specialized for specific functions. Differentiation produces a wide array of cell types, each with distinct roles, such as conducting water (xylem), transporting nutrients (phloem), and capturing light (mesophyll cells).
- ★ **External Influences on Growth:** Growth is influenced by both internal genetic factors and external environmental conditions. Light, temperature, nutrients, and water availability all play pivotal roles in determining the rate and direction of growth.
- ★ **Photoperiodism and Dormancy:** The length of daylight, known as photoperiod, influences processes like flowering. Some plants require specific day lengths to initiate flowering, while others enter periods of dormancy, a state of reduced activity, during unfavorable conditions.
- ★ **Senescence and Aging:** Over time, plants undergo senescence, a natural aging process that leads to the breakdown of cellular components. Leaves change color and fall, as the plant reallocates resources for survival during adverse conditions.

- **Reproductive Development:** Reproductive development marks a crucial phase in a plant's life cycle. Flowers form, pollination occurs, and seeds develop. This stage ensures the continuation of the species.
- **Factors Shaping Plant Architecture:** External factors, such as light intensity and direction, influence how plants grow and orient themselves. Phototropism causes plants to bend towards light sources, while gravitropism guides root growth downward.

Understanding plant growth and development is pivotal for horticulturists seeking to optimize crop yield, timing, and quality. By manipulating factors such as pruning, fertilization, and light exposure, growers can harness the power of plant growth processes to their advantage. This knowledge also serves as a foundation for plant breeding, genetic modification, and innovative agricultural practices that contribute to sustainable and resilient food production systems.

Photosynthesis and Respiration:

Photosynthesis and respiration are two intertwined processes that dictate a plant's energy balance and influence its growth, development, and overall vitality. These fundamental processes are the driving forces behind the conversion of light energy into chemical energy and the subsequent release of that energy for various cellular activities.

- **Photosynthesis:** Photosynthesis is a remarkable process through which plants harness the energy of sunlight to synthesize glucose (a form of sugar) from carbon dioxide and water. This conversion occurs within chloroplasts, specialized organelles within plant cells. Chlorophyll, the green pigment in chloroplasts, captures light energy and drives the chemical reactions that produce glucose and oxygen. The glucose serves as both an energy source for the plant and a building block for growth.
- **Respiration: Unlocking Energy Reserves:** Respiration is the counterpart to photosynthesis, representing the release of energy stored in glucose. This process takes place in all plant cells, including those that lack chloroplasts. By breaking down glucose in the presence of oxygen, plants release energy, carbon dioxide, and water. This energy fuels essential cellular activities, such as growth, reproduction, and maintenance.
- **Balance Between Processes:** The relationship between photosynthesis and respiration is critical. During daylight, photosynthesis generates

more glucose than the plant needs, allowing excess to be stored for later use. At night or when light is limited, respiration ensures a continuous supply of energy by breaking down stored glucose. This balance is crucial for maintaining plant health and vigor.

- **Environmental Impact:** Factors such as light intensity, temperature, and carbon dioxide concentration influence the rates of photosynthesis and respiration. Photosynthesis increases with adequate light and temperature, while respiration is relatively constant. This interplay between processes determines whether a plant is in a net carbon gain (photosynthesis exceeds respiration) or carbon loss (respiration exceeds photosynthesis) state.

Water and Nutrient Uptake: Nourishing Plants for Optimal Growth

Water and nutrients are essential components for sustaining plant life, growth, and overall health. The intricate processes involved in water and nutrient uptake ensure that plants receive the vital resources needed to carry out essential functions and thrive in their environments.

- **Water Uptake and Transport:** Roots play a critical role in water uptake. Through the process of osmosis, water moves from the soil into root cells. The root's structure, including root hairs, increases surface area for absorption. The vascular tissues of the plant, specifically the xylem, facilitate the upward movement of water from the roots to the shoots through capillary action and transpiration.
- **Nutrient Uptake:** Plant roots also serve as vehicles for nutrient uptake. Active transport mechanisms allow plants to absorb essential minerals from the soil. Macronutrients like nitrogen, phosphorus, and potassium, as well as micronutrients like iron and zinc, are necessary for various physiological processes, including photosynthesis and growth.
- **Root-Microbe Interactions:** Beneficial microorganisms, such as mycorrhizal fungi, form symbiotic relationships with plant roots. These fungi extend the root's reach and enhance nutrient uptake by increasing the root's surface area and accessing nutrients that are otherwise inaccessible to the plant.
- **Balancing pH and Osmosis:** The pH level of the soil affects nutrient availability. In soils with extreme pH levels, certain nutrients may

become less accessible to plants. Proper water management is essential to maintain turgor pressure within plant cells. Osmosis, the movement of water from areas of high concentration to low concentration, maintains cell structure and helps plants withstand environmental stresses.

Water and nutrient uptake are integral processes that influence plant health, growth, and productivity. By comprehending these mechanisms, horticulturists can tailor their practices to provide plants with the right amount of water and nutrients, fostering thriving and resilient crops.

Reproduction: Ensuring the Continuity of Plant Generations

Reproduction is a fundamental process that ensures the continuation of plant species by producing offspring with genetic diversity and adaptability. Plants employ various reproductive strategies, both sexual and asexual, to proliferate and thrive in diverse environments.

- **Sexual Reproduction:** Sexual reproduction involves the fusion of gametes, specialized cells from two parents, resulting in genetic diversity. In flowering plants, flowers contain reproductive structures called stamens (male) and carpels (female). Pollination transfers pollen from the stamen to the carpel, leading to fertilization and the formation of seeds. The seeds contain the genetic material necessary for the development of a new plant.
- **Fruits and Seed Dispersal:** After fertilization, the carpel develops into a fruit that protects the developing seeds. Fruits exhibit a variety of forms, textures, and mechanisms for seed dispersal. Wind, water, animals, and even explosive mechanisms aid in the dispersal of seeds away from the parent plant, increasing the chances of survival and colonization in new areas.
- **Asexual Reproduction:** Asexual reproduction involves the generation of new plants from vegetative structures of a single parent plant. This method allows plants to reproduce quickly and efficiently, but offspring are genetically identical to the parent. Common forms of asexual reproduction include runners (as in strawberries), bulbs (as in onions), and cuttings.
- **Advantages of Reproduction:** Sexual reproduction generates genetic diversity, contributing to adaptation and evolution. It also promotes

the formation of new traits that may provide advantages in changing environments. Asexual reproduction, on the other hand, ensures the preservation of favorable traits, making it useful for propagation of specific varieties.

Reproduction is a pivotal process in the life cycle of plants, guaranteeing the perpetuation of species and the ability to adapt to environmental challenges. Knowledge of reproductive strategies equips horticulturists to make informed decisions in breeding programs, conservation efforts, and the cultivation of plants for various purposes.

Plant Hormones

Plant hormones, also known as phytohormones, are chemical messengers that regulate various physiological processes, enabling plants to adapt to changing environments, coordinate growth, and respond to internal and external cues. These complex signaling molecules play a pivotal role in shaping the development and behavior of plants.

- **Auxins:** Auxins are crucial for cell elongation and growth. They are responsible for tropisms, the directional growth responses of plants towards or away from stimuli like light (phototropism) or gravity (gravitropism). Apical dominance, where the main stem inhibits the growth of lateral branches, is also influenced by auxins.
- **Gibberellins:** Gibberellins stimulate stem elongation, fruit development, and seed germination. They play a role in breaking seed dormancy and promoting flowering. Gibberellin mutants, such as dwarf varieties in agriculture, showcase their impact on plant height.
- **Cytokinins:** Cytokinins promote cell division and growth, particularly in root and shoot meristems. They interact with auxins to maintain balanced growth. Cytokinins also influence the development of lateral buds, delaying senescence and leaf abscission.
- **Abscisic Acid (ABA):** ABA is associated with stress responses and growth inhibition. It regulates stomatal closure, reducing water loss during drought conditions. ABA also promotes seed dormancy and prevents premature germination under unfavorable conditions.
- **Ethylene:** Ethylene is involved in various aspects of plant growth and development, including fruit ripening, abscission (shedding of leaves, flowers, or fruits), and response to mechanical stress or infection.

★ **Plant Hormones and Responses:** Plant hormones often work in conjunction, influencing one another's actions to achieve specific responses. For instance, auxins and gibberellins interact to stimulate stem elongation. Hormones also regulate responses to light, gravity, and stressors, allowing plants to adapt to their surroundings.

3 Soil and Soil Management

Soil is a fundamental resource in horticulture, serving as the medium through which plants obtain essential nutrients, water, and support. Successful soil management is essential for ensuring healthy growth, high yields, and quality produce in horticultural crops.

Soil Characteristics

- **Physical Properties:** Soil texture, structure, and porosity determine its physical properties. Soil texture refers to the relative proportions of sand, silt, and clay particles. Different textures influence water-holding capacity, drainage, and aeration. Soil structure relates to the arrangement of soil particles into aggregates, affecting root penetration and water movement. Proper soil structure promotes root development and prevents compaction.
- **Chemical Properties:** Soil pH, nutrient content, and cation exchange capacity (CEC) are crucial chemical properties. Soil pH influences nutrient availability; certain crops thrive in specific pH ranges. Nutrient content, including macronutrients (nitrogen, phosphorus, potassium) and micronutrients (iron, zinc, copper), impacts plant growth. CEC measures the soil's ability to hold and exchange nutrients, affecting nutrient uptake by plants.
- **Biological Properties:** Soil hosts a diverse ecosystem of microorganisms, including bacteria, fungi, and nematodes, that contribute to nutrient cycling, disease suppression, and organic matter decomposition.

Beneficial microorganisms enhance soil fertility, making soil health a critical factor in horticulture.

Importance of Soil in Horticulture

- **Nutrient Availability:** Soil serves as a reservoir of essential nutrients for plant growth. Adequate nutrient availability is critical for optimal crop development and yield. Proper soil testing guides nutrient management, preventing deficiencies or excesses that could harm plants.
- **Water Management:** Soil's water-holding capacity determines irrigation needs. Well-structured soil with good drainage prevents water logging, which can lead to root diseases. Mulching helps conserve soil moisture and regulate temperature.
- **Root Growth and Aeration:** Healthy root development relies on suitable soil structure and aeration. Compacted soil restricts root penetration and leads to stunted growth. Regular soil cultivation promotes aeration and root expansion.

Soil Management Practices

- **Soil Testing:** Regular soil testing assesses nutrient levels, pH, and other factors. Test results guide the application of fertilizers and soil amendments, ensuring crops receive the right nutrients in the right amounts.
- **Fertilization:** Understanding crop nutrient requirements aids in balanced fertilization. Organic and synthetic fertilizers offer different nutrient release rates. Applying compost or organic matter enhances soil structure, nutrient content, and microbial activity.
- **pH Adjustment:** Soil pH influences nutrient availability. Lime is used to raise pH in acidic soils, while elemental sulfur lowers pH in alkaline soils. Proper pH adjustment ensures plants can access nutrients effectively.
- **Irrigation Management:** Soil moisture monitoring helps schedule irrigation. Drip or soaker systems minimize water wastage and reduce foliar diseases by keeping foliage dry. Avoid over-irrigation to prevent water logging.

- **Mulching:** Mulching conserves soil moisture, regulates temperature, and suppresses weed growth. Organic mulches also contribute to soil organic matter content as they break down.
- **Soil Erosion Prevention:** Cover crops and contour planting prevent soil erosion, especially on sloped land. Erosion control measures safeguard topsoil and maintain soil fertility.
- **Crop Rotation and Cover Cropping:** Crop rotation breaks disease cycles and improves soil structure. Cover crops add organic matter, fix nitrogen, and protect soil during fallow periods.

Sustainable Soil Management

- **Conservation Tillage:** Reduced tillage minimizes soil disturbance, preserving soil structure and minimizing erosion. No-till practices maintain soil health while reducing the carbon footprint.
- **Integrated Pest Management (IPM):** IPM strategies reduce chemical inputs by encouraging natural pest control mechanisms, thus minimizing soil and water contamination.
- **Composting and Organic Farming:** Compost improves soil structure, water-holding capacity, and nutrient content. Organic farming practices enhance soil health by minimizing chemical inputs and promoting microbial diversity.

In horticulture, soil is a dynamic and vital resource that requires careful management to ensure sustainable crop production. Understanding soil properties, implementing appropriate management practices, and adopting sustainable approaches contribute to healthy plants, improved yields, and environmental stewardship. By prioritizing soil health, horticultural practitioners play a pivotal role in securing food production for present and future generations.

4

Water Management & Irrigation: Nurturing Plants with Adequate Hydration

Water is a precious resource that plays a vital role in sustaining plant life, promoting growth, and ensuring the health of horticultural crops. Effective water management and irrigation practices are essential for optimizing plant productivity while conserving this valuable commodity.

Importance of Water: Water is involved in numerous physiological processes within plants, including photosynthesis, nutrient uptake, and temperature regulation. Insufficient water availability can lead to wilting, reduced growth, and even death. Conversely, excessive water can cause root rot, nutrient leaching, and other detrimental effects.

Factors Influencing Irrigation: Orchestrating Water Management for Crop Success

The practice of irrigation, pivotal for sustaining agriculture and ensuring bountiful harvests, is a complex process influenced by a multitude of factors that shape how water is supplied to crops. These factors collectively determine the timing, frequency, and volume of irrigation, impacting plant health, resource efficiency, and overall productivity.

- **Soil Characteristics:** The soil's composition and structure play a crucial role in irrigation decisions. Sandy soils, with their rapid drainage, require more frequent watering to prevent drying out, while clay soils retain moisture for longer durations, allowing for less frequent irrigation. A clear understanding of the soil's water-holding capacity guides growers in tailoring their irrigation strategies to maximize water utilization.

- **Climatic Conditions:** The prevailing climate has a direct bearing on irrigation practices. In regions with high temperatures and low humidity, water evaporates rapidly, necessitating more frequent irrigation to replenish soil moisture. Conversely, areas with milder temperatures and higher humidity might require less frequent irrigation, emphasizing the need for flexibility based on local climatic variations.
- **Crop Requirements:** Different crops exhibit varying water needs based on their species, growth stages, and overall water consumption rates. Young plants, rapidly growing crops, and those with substantial foliage transpire more water and require increased irrigation. Aligning irrigation plans with the specific needs of cultivated crops is essential for sustaining healthy growth.
- **Topography and Drainage:** The physical characteristics of the land, such as its slope and drainage patterns, influence water movement and retention. Proper land grading and drainage systems prevent water logging and ensure uniform water distribution across the field. By accounting for topographical features, growers can mitigate the risk of uneven water distribution and soil erosion.
- **Irrigation Technology:** The choice of irrigation method significantly impacts water use efficiency. Drip irrigation delivers water directly to the root zone, reducing evaporation and minimizing water wastage. Sprinkler systems mimic natural rainfall, suitable for a wide range of crops. Modern technology, including soil moisture sensors and weather forecasts, enhances precision in irrigation scheduling, ensuring that water is applied when and where it is needed most.
- **Water Availability and Conservation:** The availability of water resources influences irrigation planning. Efficient water use, achieved through careful scheduling, correct application rates, and measures to minimize runoff and evaporation, is crucial for sustainable agriculture. Collecting and storing rainwater, adopting mulching practices, and optimizing irrigation timing all contribute to conserving water and reducing wastage.

Proper irrigation requires consideration of various factors, including soil type, plant species, climate, and growth stage. Soil moisture sensors and weather data can aid in determining the optimal irrigation schedule. Understanding the water needs of different crops is crucial; some prefer consistently moist conditions, while others thrive in drier environments.

- **Irrigation Methods:** Different irrigation methods cater to diverse plant requirements and local conditions. Surface irrigation involves flooding or furrow systems, while sprinkler irrigation mimics natural rainfall. Drip irrigation targets specific plant roots, minimizing water wastage. These methods impact water distribution, nutrient retention, and disease management.
- **Efficiency and Conservation:** Water efficiency and conservation are paramount. Techniques like mulching and incorporating organic matter into the soil improve water retention and reduce evaporation. Collecting and storing rainwater can supplement irrigation needs. Adopting drought-tolerant plant varieties and xeriscaping principles can minimize water consumption in landscaping.
- **Timing and Frequency:** Irrigation timing is crucial. Watering during the cooler parts of the day reduces evaporation and allows plants to absorb moisture effectively. Frequent, shallow watering can lead to shallow root growth; instead, deep and infrequent watering encourages deep root development and drought resistance.
- **Precision Irrigation:** Advancements in technology have led to precision irrigation systems that tailor water application based on plant needs. These systems use sensors, data analysis, and automation to optimize irrigation, enhancing water use efficiency.

5

Plant Nutrition & Fertilization

Plant nutrition and fertilization are foundational aspects of successful horticulture, as they directly influence plant growth, development, and overall health. Understanding the essential nutrients plants require and applying appropriate fertilization practices are essential for achieving optimal yields and sustainable agricultural practices.

Essential Nutrients: Nourishing Plants for Optimal Growth and Development

Essential nutrients are the building blocks of plant life, facilitating growth, metabolism, and overall vitality. These nutrients, required in varying quantities, play critical roles in plant functions and are pivotal for achieving healthy and productive horticultural crops.

Macronutrients:

- ✯ **Nitrogen (N):** Nitrogen is a key component of amino acids, proteins, and chlorophyll. It drives vegetative growth, enhances leaf and stem development, and influences overall plant vigor.
- ✯ **Phosphorus (P):** Phosphorus is essential for energy transfer, DNA synthesis, and root development. It promotes flowering, fruiting, and seed formation, ensuring successful reproduction.
- ✯ **Potassium (K):** Potassium regulates water movement within plants, aids in enzyme activation, and contributes to overall stress resistance. It enhances disease resistance, quality of fruits, and root health.

- **Calcium (Ca):** Calcium is essential for cell wall structure, root growth, and membrane permeability. It prevents disorders like blossom end rot in tomatoes and tip burn in lettuce.
- **Magnesium (Mg):** Magnesium is a central component of chlorophyll molecules, essential for photosynthesis. It also plays a role in enzyme activation and energy transfer.
- **Sulfur (S):** Sulfur is crucial for amino acid synthesis and protein formation. It contributes to overall plant health and supports nitrogen metabolism.

Micronutrients:

- **Iron (Fe):** Iron is essential for chlorophyll production and overall photosynthesis. It aids in electron transfer and is vital for enzyme function.
- **Zinc (Zn):** Zinc plays a role in enzyme activity, hormone regulation, and protein synthesis. It influences plant growth and development.
- **Manganese (Mn):** Manganese is involved in photosynthesis, nitrogen metabolism, and the breakdown of carbohydrates. It supports root growth and overall plant health.
- **Copper (Cu):** Copper is crucial for enzyme activity, particularly in reactions related to cell wall development and lignin formation.
- **Molybdenum (Mo):** Molybdenum is essential for nitrogen fixation and the conversion of nitrates to ammonia in plants.
- **Boron (B):** Boron is involved in cell wall formation, flower development, and calcium uptake. It influences overall plant structure and reproductive processes.
- **Chlorine (Cl):** While needed in very small quantities, chlorine supports photosynthesis and osmotic balance in plants.

Nutrient Uptake: Navigating the Pathways of Plant Nutrition

The process of nutrient uptake in plants is a complex interplay of biological mechanisms that ensure essential elements are absorbed from the soil and transported to various parts of the plant. This intricate process is essential for plant growth, development, and overall health.

- **Root Structure and Function:** Plant roots are the primary organs responsible for nutrient uptake. Root hairs, fine extensions of root cells, greatly increase the surface area available for nutrient absorption. The root's epidermis, cortex, and endodermis layers are involved in regulating nutrient flow.
- **Passive and Active Transport:** Nutrient uptake involves both passive and active transport mechanisms. Passive transport relies on diffusion, where nutrients move from areas of higher concentration in the soil to areas of lower concentration in the root. Active transport requires energy and specialized transport proteins to move nutrients against their concentration gradients.
- **Root Exudates and Symbiotic Relationships:** Root exudates, compounds released by roots, influence soil microbial communities and nutrient availability. Beneficial microorganisms, such as mycorrhizal fungi, form symbiotic relationships with plant roots, extending the root's reach and enhancing nutrient uptake.
- **Nutrient Mobility and Mobility Factors:** Nutrients vary in their mobility within plants. Mobile nutrients, like nitrogen, can move from older to younger leaves, ensuring essential processes like photosynthesis are sustained. Immobile nutrients, such as calcium, remain in the tissues where they were initially deposited.
- **Nutrient Translocation:** Once absorbed, nutrients are transported throughout the plant via the xylem and phloem. The xylem carries water and nutrients from the roots to the shoots, while the phloem transports sugars and other organic compounds to various parts of the plant.
- **Influences on Nutrient Uptake:** Several factors impact nutrient uptake. Soil pH affects nutrient solubility, influencing their availability for absorption. Soil moisture content, temperature, and root health also play roles in nutrient uptake efficiency.
- **Nutrient Competition and Antagonism:** Nutrient uptake is influenced by interactions between different elements. Nutrient competition can occur when one nutrient's presence affects the absorption of another. Nutrient antagonism arises when an excess of one nutrient interferes with the uptake of another.

Fertilization Strategies: Nourishing Plants for Maximum Growth and Yield

Fertilization is a critical practice in horticulture that involves supplying essential nutrients to plants to overcome deficiencies and enhance growth. Effective fertilization strategies are designed to optimize nutrient availability, promote healthy development, and ensure the success of horticultural crops.

- ✯ **Understanding Nutrient Needs:** Before implementing fertilization strategies, it's essential to determine the nutrient requirements of specific crops. Soil testing helps identify nutrient deficiencies or imbalances, guiding the selection of appropriate fertilizers and application rates.
- ✯ **Macronutrient and Micronutrient Considerations:** Fertilizers are formulated to address the nutrient needs of plants. Macronutrient fertilizers, such as those with N-P-K ratios, provide primary elements like nitrogen, phosphorus, and potassium. Micronutrient fertilizers contain trace elements like iron, zinc, and copper, addressing specific deficiencies.
- ✯ **Organic vs. Synthetic Fertilizers:** Fertilizers can be organic, derived from natural sources like compost, manure, or plant residues, or synthetic, chemically formulated to contain specific nutrients. Organic fertilizers release nutrients gradually as they decompose, while synthetic fertilizers provide a quick nutrient boost.

Application Methods

- ✯ **Broadcasting:** Evenly spreading fertilizers across the entire area is a common method. It's suitable for larger fields but may result in nutrient wastage.
- ✯ **Top-Dressing:** Applying fertilizers around the base of plants targets nutrient delivery to the root zone, minimizing contact with leaves.
- ✯ **Incorporation:** Fertilizers can be mixed into the soil before planting, ensuring nutrients are available as plants grow.
- ✯ **Foliar Spraying:** Nutrient solutions can be sprayed directly onto plant leaves, allowing for rapid absorption. This method is particularly useful for correcting micronutrient deficiencies.
- ✯ **Timing and Frequency:** Proper timing of fertilizer application is crucial. Pre-planting fertilization provides nutrients for initial growth, while side-dressing during the growing season replenishes nutrients.

Applying fertilizers before heavy rain can lead to nutrient runoff, while late-season fertilization may encourage vegetative growth instead of fruiting.

- **Slow-Release Fertilizers:** Slow-release fertilizers provide nutrients gradually over an extended period, reducing the risk of over-fertilization and minimizing nutrient loss through leaching.
- **Organic Matter and Cover Crops:** Incorporating organic matter into the soil enhances nutrient availability and improves soil structure. Cover crops can also add nutrients and improve soil health through their decomposition.
- **Balancing Nutrients:** It's crucial to strike a balance between different nutrients. Over-fertilization can lead to nutrient imbalances, affecting plant health and yield.
- **N-P-K Ratio:** The N-P-K ratio on fertilizer labels indicates the proportion of nitrogen (N), phosphorus (P), and potassium (K) in the fertilizer. This ratio aligns with the specific growth stages and needs of plants. Nitrogen promotes leafy growth, phosphorus supports root development and flowering, and potassium enhances overall plant health and stress tolerance.
- **Soil Testing:** Soil testing is a fundamental tool for determining nutrient levels in the soil. By understanding existing nutrient deficiencies or excesses, growers can tailor their fertilization approach to address specific needs, avoiding wastage and environmental harm.
- **Nutrient Imbalances and Toxicities:** Over-fertilization can lead to nutrient imbalances and toxicity, negatively impacting plant health and the environment. Careful application of fertilizers, guided by soil test results and plant observations, prevents these issues.
- **Sustainable Practices:** Adopting sustainable fertilization practices is crucial for long-term soil health and environmental preservation. Balanced fertilization, integrating organic matter, and practicing crop rotation contribute to the prevention of nutrient depletion and soil degradation.

Plant nutrition and fertilization are at the core of healthy crop production. By supplying plants with the nutrients they need in the right amounts and utilizing sustainable fertilization practices, horticulturists can enhance plant vigor, increase yields, and contribute to the overall sustainability of agricultural systems.

6
Plant Propagation

Plant propagation is the process of creating new plants from existing ones. It's a fundamental practice in horticulture, allowing growers to reproduce plants with desirable traits and expand their plant stock.

Sexual Propagation

Seed Propagation:

Seeds are the product of sexual reproduction and offer a diverse range of genetic characteristics. This technique involves sowing seeds in a suitable growing medium and providing proper conditions for germination. Seed propagation is commonly used for annuals, perennials, and some trees.

Cross-Pollination and Hybridization

Controlled cross-pollination between different plant varieties can produce hybrids with desired traits. Hybridization is crucial in developing new cultivars that possess improved characteristics, such as disease resistance or better fruit quality.

Asexual Propagation

Cutting Propagation

Stem or leaf cuttings can develop into new plants. Techniques include hardwood, softwood, and semi-hardwood cuttings. Rooting hormones can enhance success rates. This method is widely used for ornamental plants and woody species.

Layering

Layering involves inducing roots to form on a stem while it is still attached to the parent plant. Air layering and simple layering are common methods used to propagate plants like fruit trees and shrubs.

Division

Plants with clumping growth habits, such as ornamental grasses and perennials, can be divided into smaller sections, each of which grows into an independent plant. This technique also helps rejuvenate older plants.

Grafting and Budding

Grafting involves joining the cut surfaces of two plants to grow as one. Budding is a specific form of grafting where a bud from the desired plant (the scion) is inserted into a cut on the rootstock. These techniques are used for fruit trees, ornamental plants, and to propagate rare species.

Tissue Culture and Micropropagation

Tissue culture and micropropagation are techniques used in plant propagation, where small pieces of plant tissues are cultured in a controlled environment to produce new plants. This process is particularly useful for rapidly producing large numbers of plants with desirable traits, such as disease resistance, uniform growth, and genetic stability. Here are the general steps involved in tissue culture and micropropagation:

- **Initiation and Selection of Explant:** An explant is a small piece of plant tissue, usually taken from the shoot tip or the meristem (growing point). The explant should be free from diseases and contaminants. It's selected based on the type of plant and the desired outcome.
- **Surface Sterilization:** The explant is subjected to a sterilization process to eliminate any surface contaminants, such as bacteria, fungi, or viruses. This is usually done by dipping the explant in a sterilizing solution, often containing a mixture of bleach and detergent.
- **Establishment of Aseptic Culture:** The sterilized explant is placed in a sterile culture medium containing nutrients, vitamins, growth

regulators (such as auxins and cytokinins), and sugars. The culture medium provides the necessary nutrients for the explant to grow.

- **Shoot Initiation:** The explant is encouraged to produce shoots by manipulating the concentrations of growth regulators in the culture medium. Cytokinins promote shoot formation, while auxins control root development. The goal is to produce multiple shoots from a single explant.
- **Shoot Multiplication:** The developed shoots are separated and transferred to fresh culture media to promote their multiplication. This involves periodically subculturing the shoots onto new media to prevent overcrowding and to encourage healthy growth.
- **Root Induction:** To facilitate the transition from the in vitro (lab) environment to the ex vitro (natural) environment, the developed shoots are subjected to a medium with higher levels of auxins to stimulate root formation.
- **Acclimatization:** Once roots are established, the plant lets are gradually exposed to external environmental conditions. This process helps the plant lets adapt to the real-world conditions outside the controlled lab environment. Initially, they might be placed in a humidity chamber or under controlled conditions before being transferred to the field or greenhouse.
- **Transplanting:** Once the plant lets have adapted and grown to a sufficient size, they can be transplanted into the desired location, such as a field or a greenhouse, where they will continue to grow and develop into mature plants.

Specialized Techniques

- **Bulb Division:** Bulbs, corms, and tubers are underground storage structures that can be divided to create new plants. This method is common for plants like daffodils and irises.
- **Rhizome Division:** Rhizomes are horizontal underground stems that can be divided to create new plants. Bamboo and ginger are commonly propagated using this method.
- **Suckering and Offsets:** Plants that produce suckers or offsets, like strawberries and aloe vera, can be separated from the parent plant and replanted.

Factors Influencing Propagation Success

The success of plant propagation, including techniques like tissue culture and micro propagation, can be influenced by a variety of factors. These factors can impact the growth and development of plants at different stages of the propagation process. Here are some key factors that can influence propagation success:

- **Genetic Factors**: The genetic makeup of the plant material being propagated plays a significant role. Some plant varieties or species might be more amenable to propagation than others. Traits like growth rate, disease resistance, and overall vigor can affect success.
- **Plant Health and Quality:** The initial health and quality of the donor plant material (explants) are crucial. Plants that are free from diseases, pests, and physiological disorders are more likely to propagate successfully.
- **Explants Selection:** The choice of explants, whether they are shoot tips, axillary buds, or other plant parts, can impact success. The physiological state of the explant (young and actively growing tissue) is important for successful initiation.
- **Sterilization:** Proper sterilization of the explants is essential to prevent contamination by bacteria, fungi, and other microorganisms that could inhibit growth.
- **Culture Medium:** The composition of the culture medium, including nutrients, vitamins, growth regulators, and sugars, affects the growth and development of the explants. The concentrations of these components need to be carefully optimized for each plant species.
- **Aseptic Techniques:** Maintaining aseptic (sterile) conditions throughout the propagation process is critical to prevent contamination and ensure healthy growth.
- **Environmental Conditions:** Environmental factors such as temperature, humidity, light intensity, and photoperiod (light duration) in the growth chamber or greenhouse can influence plant growth and development.
- **pH and Water Quality:** The pH of the culture medium and the quality of water used for irrigation can impact nutrient availability and plant health.
- **Microbial Contamination:** Bacterial, fungal, and viral contamination can negatively impact propagation success. Regular monitoring and appropriate antimicrobial strategies are essential.

- ✭ **Subculturing and Multiplication:** Regular subculturing of the explants onto fresh culture media is necessary to prevent overcrowding, which can lead to reduced growth and health.
- ✭ **Rooting Hormones:** The type and concentration of rooting hormones (auxins) used during root induction can influence the success of root formation and subsequent acclimatization.
- ✭ **Acclimatization:** Proper hardening-off procedures during acclimatization help the plantlets adapt to external environmental conditions. Gradually exposing them to changes in temperature, humidity, and light can improve survival rates.
- ✭ **Handling and Care:** Gentle handling of plant materials during all stages of propagation minimizes stress and damage, leading to better outcomes.
- ✭ **Disease and Pest Management:** Monitoring for diseases and pests during the propagation process and taking appropriate measures to manage them are essential to prevent losses.
- ✭ **Nutrient Management:** Ensuring proper nutrient levels and ratios in the culture medium is important for healthy growth and development.
- ✭ **Ethical and Legal Considerations:** Some plant species may have legal restrictions on propagation due to conservation concerns or intellectual property rights.

It's important to recognize that different plants may have specific requirements, and propagation techniques may need to be adjusted accordingly. Additionally, advancements in technology and research may lead to new insights and methods for improving propagation success.

7

Importance of Rootstocks in Horticultural Crops

Introduction

Rootstocks are an essential component of modern horticulture, contributing significantly to the success, productivity, and sustainability of various fruit-bearing plants. Rootstocks serve as the foundation upon which scion cultivars are grafted, imparting a range of attributes that influence plant growth, development, and overall performance. This chapter explores the profound importance of rootstocks in horticultural crops, shedding light on their impact on crop yield, disease resistance, abiotic stress tolerance, and more.

Enhancing Disease Resistance

One of the primary reasons for using rootstocks is to confer disease resistance to scion cultivars. Many fruit crops are susceptible to soilborne pathogens and nematodes. Rootstocks with innate resistance to these pests provide a shield against diseases that could otherwise devastate susceptible cultivars. For example, citrus rootstocks resistant to citrus tristeza virus have been instrumental in ensuring the viability of citrus production.

Abiotic Stress Tolerance

Rootstocks play a critical role in imparting abiotic stress tolerance to horticultural crops. They can enhance plants' ability to withstand adverse conditions such as salinity, drought, and temperature extremes. By developing rootstocks that possess these stress-tolerant traits, growers can expand their cultivation areas and maintain consistent yields even in challenging environments.

Improving Soil Adaptability

Different rootstocks exhibit varying degrees of adaptability to different soil types and conditions. Some rootstocks have the capacity to thrive in soils with poor drainage, high pH, or other unfavorable characteristics. By selecting the appropriate rootstock for the specific soil conditions of a given location, growers can optimize plant health and yield.

Managing Growth Habits

Rootstocks also play a role in controlling plant growth habits. Dwarfing rootstocks are used to limit the size of fruit trees, making them more manageable and easier to harvest. This is particularly relevant in urban or high-density planting systems. Conversely, vigorous rootstocks can be chosen to encourage larger trees with higher yields.

Compatibility with Scion Cultivars

Choosing the right rootstock is essential for ensuring compatibility with the scion cultivar. An incompatible rootstock can lead to poor growth, reduced fruit quality, and decreased overall performance. Proper matching of rootstock and scion characteristics is a fundamental consideration in successful grafting and horticultural crop production.

Research and Breeding Efforts

Ongoing research and breeding efforts are focused on developing rootstocks with enhanced traits, such as disease resistance, stress tolerance, and improved fruit quality. This reflects the continuous pursuit of innovation to address the evolving challenges faced by horticultural growers worldwide.

Implications and Applications

The use of rootstocks in horticultural crops exemplifies the intricate synergy between genetics and cultivation practices. By strategically selecting and grafting appropriate rootstocks, growers can optimize plant health, enhance disease resistance, mitigate abiotic stress, and ultimately achieve more productive and sustainable horticultural systems.

Disease Resistance

Certain rootstocks possess inherent resistance to soilborne pathogens, nematodes, and fungal diseases. By grafting a scion onto a disease-resistant rootstock, farmers can mitigate the negative impact of soilborne pathogens on crop health.

Abiotic Stress Tolerance

Rootstocks can confer tolerance to abiotic stressors such as salinity, drought, and temperature extremes. This is particularly important in regions with challenging environmental conditions.

Nutrient Uptake Efficiency

Rootstocks with efficient nutrient uptake capabilities can enhance overall plant nutrition, leading to increased crop yield and quality.

Vigor and Yield Improvement

Grafted plants often exhibit increased vigor, resulting in improved growth, larger fruit sizes, and enhanced crop yield.

Grafting Techniques

Grafting involves the precise joining of the scion and rootstock to create a unified plant. The techniques used for grafting vary depending on the plant species and desired outcome:

- **Cleft Grafting:** In cleft grafting, a vertical slit is made in the rootstock, and a scion is inserted into the cleft. This technique is commonly used for grafting tomato and other solanaceous crops.
- **Whip-and-Tongue Grafting:** Whip-and-tongue grafting involves creating matching diagonal cuts on both the scion and rootstock, which are then interlocked. This technique is widely used for small seedlings and precise alignment.
- **Side Grafting:** In side grafting, a lateral incision is made on the rootstock stem, and the scion is inserted into the incision. This technique is suitable for large seedlings and trees.

Practical Applications

The use of rootstocks in vegetable production has become a standard practice in various cropping systems:

- **Tomato Production:** Grafted tomato plants are commonly used to enhance disease resistance, extend the growing season, and improve overall crop yield and quality.
- **Cucurbit Production:** Grafting cucurbit crops such as melons, cucumbers, and pumpkins onto disease-resistant rootstocks can prevent the occurrence of soilborne diseases and increase yield.
- **Solanaceous Crop Production:** Pepper and eggplant crops also benefit from grafting, as it enhances resistance to soil borne pathogens and can improve crop performance in marginal soils.

8

Principles of Pest & Disease Management in Horticultural Crops

The cultivation of horticultural crops is often challenged by the presence of pests and diseases that can significantly impact yield and quality. Effective pest and disease management is essential to ensure sustainable and productive horticultural systems. This chapter outlines the key principles and strategies for managing and controlling pests and diseases in horticultural crops, focusing on integrated approaches that minimize environmental impact and promote long-term success.

Cultural Practices for Prevention

Cultural practices are fundamental techniques used in horticultural crop management to prevent and mitigate the impact of pests and diseases. By creating an unfavorable environment for these organisms, cultural practices contribute to the overall health and productivity of crops while minimizing the need for chemical interventions. Here are some key cultural practices for pest and disease prevention in horticultural crops:

- **Crop Rotation:** Crop rotation involves changing the type of crop grown in a specific field from season to season. This practice disrupts the life cycles of pests and diseases that are specific to certain crops. Different crops have varying nutrient requirements, which can help break the cycle of soil-borne pathogens and pests that rely on continuous cultivation of their host plants.

- **Proper Plant Spacing:** Adequate plant spacing ensures good air circulation and sunlight penetration among plants. Proper spacing reduces humidity levels, which can discourage the development and spread of fungal diseases. It also makes it more difficult for pests to move from plant to plant.
- **Sanitation and Removal of Debris:** Regularly removing plant debris, fallen leaves, and weeds from the growing area eliminates potential hiding places and breeding grounds for pests and pathogens. These materials can harbor pests and diseases over the offseason, allowing them to resurface and infect new crops in the next planting season.
- **Soil Health Management:** Healthy soil supports vigorous plant growth and helps plants resist pests and diseases. Practices such as proper irrigation, balanced fertilization, and maintaining appropriate soil pH promote strong plant development and reduce stress that can make plants more susceptible to pests and diseases.
- **Selection of Disease-Resistant Varieties:** Choosing crop varieties that are naturally resistant to specific pests and diseases can greatly reduce the risk of infestations. Plant breeding programs have developed varieties with genetic traits that make them less susceptible to certain pathogens and pests.
- **Polyculture and Companion Planting:** Growing a diverse mix of crops in the same area, known as polyculture, can disrupt the habitat of pests and reduce their ability to establish large populations. Companion planting involves strategically placing plants that repel pests or attract beneficial insects next to susceptible crops.
- **Timing of Planting and Harvesting:** Proper timing of planting and harvesting can reduce the exposure of crops to peak periods of pest and disease activity. Early planting can help crops establish before pests become abundant, while timely harvesting prevents overripe or decaying plant material that can attract pests.
- **Use of Clean Plant Material:** Starting with clean, disease-free plant material is essential to preventing the introduction of pests and diseases. Quarantining new plants before introducing them to the growing area allows for inspection and treatment if necessary.
- **Mulching:** Applying organic mulch around the base of plants helps regulate soil temperature, conserve moisture, and suppress weed growth. Mulch can also act as a physical barrier that prevents soil-borne pathogens from splashing onto plant foliage during rain events.

* **Monitoring and Record Keeping:** Regular monitoring of crops for signs of pests and diseases enables early detection and intervention. Keeping accurate records of pest and disease occurrences, weather conditions, and management practices can help identify trends and refine preventive strategies over time.

Cultural practices form the foundation of pest and disease prevention in horticultural crops. By implementing these practices, growers can create a healthier growing environment that reduces the reliance on chemical treatments and fosters sustainable and resilient crop systems. Customizing these practices to the specific needs of each crop and local conditions enhances their effectiveness in maintaining the long-term health and productivity of horticultural crops.

Biological Control

Biological control is a sustainable and environmentally friendly approach used in horticultural crops to manage pests and diseases. By capitalizing on the natural relationships between organisms, biological control utilizes predators, parasitoids, and pathogens to regulate pest populations. These natural enemies target specific pests, helping to maintain a balanced ecosystem and reduce the need for chemical pesticides.

Predators, such as ladybugs and predatory mites, consume pest insects, while parasitoids lay eggs on or inside host pests, eventually killing them. Pathogens, such as fungi and viruses, infect pests, leading to their demise. These beneficial organisms are often introduced or encouraged in the growing area to control pests effectively. Biological control not only offers long-term solutions to pest problems but also contributes to biodiversity conservation and the preservation of natural processes. When integrated into an overall pest management strategy, biological control enhances the resilience of horticultural systems while minimizing the environmental impact associated with chemical interventions.

Chemical Management

Chemical management in horticultural crops involves the strategic use of chemical pesticides to control pests and diseases. While these chemicals can provide rapid and targeted solutions, their application requires careful consideration to mitigate potential environmental and health impacts.

Responsible chemical management prioritizes the judicious and precise use of pesticides to achieve effective pest control while minimizing risks.

One key aspect of responsible chemical management is the selection of appropriate pesticides that target specific pests or diseases while causing minimal harm to non-target organisms and the environment. Integrated Pest Management (IPM) principles advocate for using chemicals as part of a broader strategy that includes cultural practices, biological controls, and monitoring. This approach reduces the overall reliance on chemical treatments and promotes the sustainability of horticultural systems. By adhering to recommended application rates, proper timing, and techniques that minimize drift and runoff, growers can ensure that chemical management remains an effective tool in their pest management toolbox while upholding environmental stewardship.

Monitoring and Early Detection: Key Pillars of Effective Pest Management

Monitoring and early detection form the foundation of successful pest management strategies in horticultural crops. Regular and systematic monitoring involves consistent observations of plant health, pest populations, and disease symptoms. This proactive approach enables growers to detect potential issues before they escalate into significant problems, allowing for timely interventions.

Early detection is crucial because it enables growers to take swift action to prevent the rapid spread of pests and diseases. By identifying pests or pathogens at their initial stages, growers can employ targeted interventions that are less resource-intensive and have a lower environmental impact compared to tackling larger infestations. Various monitoring methods, such as visual inspections, sticky traps, and remote sensing technologies, contribute to a comprehensive understanding of the pest and disease dynamics within the crop environment. When integrated into a larger pest management strategy, monitoring and early detection empower growers to make informed decisions, optimize resource allocation, and ultimately achieve healthier and more productive horticultural systems.

Host Plant Resistance and Breeding

Host plant resistance involves developing horticultural crop varieties that possess inherent tolerance or immunity to specific pests and diseases. Plant breeders identify and incorporate resistance genes from wild relatives or related species into cultivated varieties, creating plants with enhanced defense mechanisms. This approach reduces the need for chemical interventions and contributes to sustainable pest and disease management. Breeding programs also consider other desirable traits such as yield, quality, and adaptation to local conditions, ensuring that resistant varieties meet the needs of growers and consumers. By deploying these resistant cultivars, growers establish a frontline defense against pests and diseases, promoting long-term sustainability in horticultural production systems. As a result, host plant resistance and breeding serve as a valuable tool in integrated pest management, enhancing crop resilience and reducing environmental impact.

Quarantine and Exclusion Measures

Quarantine and exclusion measures are critical components of pest and disease management in horticultural crops, particularly in preventing the introduction and spread of new or exotic pests and diseases. Quarantine involves stringent regulations and inspection procedures for incoming plant materials, ensuring that they are free from harmful pests and pathogens before entering a new area. This proactive approach helps prevent the establishment of potentially devastating organisms in new environments, safeguarding local agriculture and ecosystems, focus on creating physical barriers to limit the access of pests and diseases to growing areas. Techniques such as installing nets, screens, or fences can prevent the entry of pests, thereby reducing the need for chemical interventions. Additionally, these measures contribute to biosecurity by controlling the movement of pests between different regions or countries. Overall, the combination of quarantine and exclusion strategies forms a vital frontline defense, reducing the risk of pest and disease outbreaks and supporting the long-term health and sustainability of horticultural systems.

Integrated Pest Management (IPM)

Integrated Pest Management (IPM) stands as a cornerstone of modern horticultural practices, providing a comprehensive and environmentally conscious framework for managing pests and diseases. Rooted in the principle of balance, IPM integrates various pest control strategies to minimize the use of chemical pesticides while maintaining healthy crops and ecosystems. At its core, IPM seeks to prevent pest and disease issues through a combination of strategies. Cultural practices, such as crop rotation, proper sanitation, and the use of disease-resistant varieties, establish unfavorable conditions for pests to thrive. Regular monitoring of pest populations allows growers to identify potential problems early, setting action thresholds that trigger interventions only when pest numbers reach a certain level. This approach prevents excessive use of chemical pesticides and the subsequent risks to human health, non-target organisms, and the environment.

IPM's strength lies in its integrated strategies. Biological control involves introducing natural enemies to prey upon or parasitize pests, reducing their numbers naturally. Additionally, the targeted use of selective pesticides, chosen for minimal impact on non-target species, is employed judiciously when other measures are insufficient. Physical barriers, like nets or traps, can also prevent pests from reaching crops. This multifaceted approach ensures a balance between controlling pests and maintaining the health of the agro ecosystem. The benefits of IPM are manifold. Its environmentally responsible practices minimize chemical residues in food and soil, preserving ecosystem integrity and human health. By addressing the underlying causes of pest problems, IPM provides more sustainable, long-term solutions compared to reliance on chemicals alone. Moreover, it aids in the management of pest resistance, as the diversity of strategies reduces the selective pressure on pests to develop resistance to a single control method.

However, implementing IPM demands knowledge, commitment, and continuous learning. Growers need to understand pest behavior, biological interactions, and local conditions to tailor IPM strategies effectively. This requires collaboration between growers, researchers, and extension services to share insights, monitor progress, and refine practices over time. Despite the challenges, IPM remains a pivotal approach in ensuring food security, economic viability, and environmental health in modern horticultural systems.

IPM combines multiple strategies into a holistic approach. It considers factors such as pest biology, crop phenology, and environmental conditions to devise a customized plan. By synergizing cultural, biological, and chemical methods, IPM reduces the risk of resistance and enhances overall effectiveness.

Education and Extension

Education and extension services play a crucial role in disseminating knowledge, building skills, and fostering best practices in pest and disease management among horticultural growers. These services bridge the gap between scientific research and on-the-ground implementation, empowering farmers to make informed decisions that contribute to the sustainability and productivity of their operations.

Effective education and extension programs provide growers with up-to-date information on the latest pest control techniques, integrated approaches, and emerging threats. Workshops, seminars, and training sessions allow farmers to learn about the biology of pests, disease cycles, and the principles of integrated pest management (IPM). By understanding these concepts, growers can implement proactive measures, make informed choices about interventions, and adopt preventive practices that reduce the reliance on chemical treatments.

Furthermore, extension services offer personalized guidance, tailoring recommendations to local conditions, crop types, and individual needs. Extension agents often conduct on-site visits to assess pest and disease situations, answer questions, and provide hands-on demonstrations of IPM techniques. By building strong relationships with growers, extension services foster a culture of continuous learning and improvement. Overall, education and extension form a dynamic partnership that equips horticultural practitioners with the knowledge and tools needed to address pest and disease challenges effectively while promoting environmentally sustainable practices.

Sustainable Practices and Environmental Concerns

In the pursuit of productive horticultural systems, the integration of sustainable practices is paramount to ensure both short-term success and long-term environmental health. Sustainable practices prioritize the careful management of resources, reduction of negative impacts, and preservation of biodiversity. By minimizing the ecological footprint of horticultural activities, growers contribute to the overall resilience of agricultural ecosystems.

However, as horticultural systems expand to meet global demands, environmental concerns come to the forefront. Excessive use of chemical pesticides and fertilizers can lead to soil degradation, water pollution, and harm to non-target organisms. Sustainable practices advocate for reduced chemical inputs through approaches like integrated pest management (IPM) and organic farming. Additionally, responsible water management, efficient irrigation techniques, and soil conservation practices conserve natural resources and prevent unnecessary waste. Biodiversity is also a cornerstone of sustainability, as it promotes natural pest control and enhances ecosystem stability. Embracing agroecological principles, such as incorporating cover crops and hedgerows, supports beneficial organisms while maintaining a balanced agroecosystem. By addressing these environmental concerns and adopting sustainable practices, horticultural growers contribute to the preservation of ecosystems, safeguarding not only their livelihoods but also the well-being of future generations.

9

Horticultural Tools and Equipments

Horticultural tools and equipment are essential for various tasks involved in gardening, landscaping, plant propagation, and other horticultural activities. These tools help horticulturists and gardeners maintain, cultivate, and care for plants effectively. Here's a detailed list of common horticultural tools and equipment:

1. Hand Tools:

- **Pruning Shears:** Used for trimming and pruning small branches and stems.
- **Secateurs:** Similar to pruning shears, used for precise cutting of small branches and stems.
- **Hedge Shears:** Designed for shaping and trimming hedges and shrubs.
- **Loppers:** Used for cutting thicker branches and stems that are beyond the capacity of pruning shears.
- **Hand Trowel:** Small shovel-like tool for digging holes, transplanting, and soil work.
- **Hand Fork:** Used for loosening soil, removing weeds, and aerating small areas.
- **Weeder:** Designed to remove weeds from the root without disturbing surrounding plants.
- **Hand Cultivator:** Used to break up soil, remove weeds, and cultivate the top layer of the soil.

- ★ **Hand Rake:** Used for smoothing soil, leveling, and collecting debris.
- ★ **Transplanting Spade:** Smaller and narrower than a regular shovel, used for planting and digging in tight spaces.
- ★ **Bypass Pruner:** Ideal for precise cuts on live plants, preventing crushing of stems.

2. Digging and Planting Tools:

- ★ **Shovel:** Used for digging, lifting, and moving soil and other materials.
- ★ **Spade:** Similar to a shovel but with a flat blade, ideal for cutting through roots and edging.
- ★ **Fork Spade:** Features fork-like tines for digging and aerating soil.
- ★ **Post Hole Digger:** Used to dig holes for planting posts, trees, and larger plants.
- ★ **Bulb Planter:** Designed to create holes for planting bulbs at a consistent depth.
- ★ **Dibble Bar:** A pointed metal bar used to create holes in soil for planting.

3. Soil Preparation Tools:

- ★ **Rototiller:** A motorized tool used to till and prepare soil for planting on a larger scale.
- ★ **Cultivator:** Hand held or motorized tool used for tilling and cultivating soil.
- ★ **Soil Knife:** Versatile tool for various tasks including planting, weeding, cutting roots, and more.
- ★ **Mattock:** Combination tool with a pick-like end for breaking up soil and a flat blade for digging.

4. Watering Tools:

- ★ **Watering Can:** Used to manually water plants with a gentle flow of water.
- ★ **Hose and Nozzle:** A hosepipe with a nozzle attachment for controlled watering.
- ★ **Sprinkler:** Distributes water over a large area, ideal for lawns and flower beds.
- ★ **Drip Irrigation Kit:** Provides a controlled and efficient way to water plants at the root zone.

5. Pruning and Cutting Tools:

- ✯ **Pruning Saw:** Designed for cutting thicker branches and stems during pruning.
- ✯ **Bow Saw:** Used for cutting larger branches and firewood.
- ✯ **Hatchet:** A small, versatile tool for chopping and splitting wood.
- ✯ **Grafting Knife:** Specialized knife for grafting and budding techniques in plant propagation.
- ✯ **Scissors:** Used for precision cutting, deadheading, and delicate pruning.

6. Measuring and Layout Tools:

- ✯ **Tape Measure:** Used for measuring distances, dimensions, and spacing.
- ✯ **Ruler:** Handy for smaller measurements and layout tasks.
- ✯ **Stakes and String:** Used to create straight lines, mark boundaries, and guide planting.

7. Protective Gear:

- ✯ **Gloves:** Protect hands from thorns, rough surfaces, and chemicals.
- ✯ **Safety Goggles:** Protect eyes from debris and flying particles.
- ✯ **Hat and Sunscreen:** Shield from the sun's rays during outdoor work.

8. Miscellaneous Tools:

- ✯ **Wheelbarrow:** Used for transporting soil, plants, and other materials.
- ✯ **Garden Cart:** Similar to a wheelbarrow but with four wheels, ideal for heavier loads.
- ✯ **Leaf Rake:** Designed for collecting leaves, grass clippings, and other debris.
- ✯ **Leaf Blower:** Used to blow leaves and debris from surfaces.
- ✯ **Compost Bin:** Container for composting organic materials to create nutrient-rich soil amendment.
- ✯ **pH Tester:** Measures the pH level of the soil to ensure optimal plant growth.
- ✯ **Sprayers:** Used for applying fertilizers, pesticides, and other solutions.

It's important to choose the right tool for each task, keep them clean and well-maintained, and use them safely to achieve the best results in your horticultural activities.

10
Greenhouse and Nursery Management

Greenhouse: A greenhouse is a controlled environment structure designed for the cultivation of plants. It allows gardeners, horticulturists, and farmers to extend the growing season, protect plants from adverse weather conditions, and create an optimal environment for plant growth by regulating factors like temperature, humidity, light, and ventilation.

Types of Greenhouses

1. Traditional Greenhouses

These are the standard type of greenhouses, often made of glass or plastic panels. They provide a controlled environment for a wide range of plants and can be customized with heating, cooling, and ventilation systems.

2. Cold Frames

Cold frames are low, enclosed structures with a transparent cover that helps protect plants from frost and cold temperatures. They are often used to harden off seedlings before transplanting them outdoors.

3. High Tunnels

Also known as hoop houses, high tunnels are simple greenhouse structures made with metal or plastic hoops covered with plastic film. They provide some protection from weather extremes and are less expensive than traditional greenhouses.

4. Polytunnels

Similar to high tunnels, polytunnels are large, curved structures covered with polyethylene plastic. They are often used for commercial crop production.

5. Shade Houses

These structures are designed to provide shade to plants and protect them from intense sunlight. They are commonly used for growing shade-loving plants and for acclimatizing plants before transplanting.

6. Aquaponic Greenhouses

These combine plant cultivation with aquaculture (fish farming). Nutrient-rich water from fish tanks is used to fertilize plants, while plants help filter and clean the water for the fish.

7. Multi-Span Greenhouses

These are large greenhouses composed of multiple connected sections (spans). They allow for efficient space utilization and are often used for commercial crop production.

Nursery Management

Nursery management involves the cultivation and propagation of plants in controlled environments, with the goal of producing healthy and vigorous plants for sale or transplanting. Here are the key aspects of nursery management:

1. Site Selection

Choose a location with suitable soil, drainage, access to water, and favorable climatic conditions for the types of plants you plan to grow.

2. Infrastructure

Set up facilities including greenhouses, shade houses, irrigation systems, potting areas, and storage spaces.

3. Propagation

Use techniques like seed sowing, cuttings, grafting, and tissue culture to produce plants. Maintain optimal conditions for germination and growth.

4. Plant Care

Provide appropriate nutrition, water, and care to ensure healthy growth. Regularly monitor for pests, diseases, and nutrient deficiencies.

5. Transplanting

Pot up or transplant seedlings into appropriate containers or beds when they reach the desired size.

6. Harden Off

Gradually expose plants to outdoor conditions to acclimate them before sale or transplanting.

7. Record Keeping

Maintain accurate records of plant varieties, propagation methods, growth time lines, and any treatments applied.

8. Marketing and Sales

Develop marketing strategies to promote and sell plants to customers, including individuals, landscapers, garden centers, and other businesses.

9. Quality Control

Implement quality control measures to ensure that only healthy and pest-free plants are sold.

10. Environmental Consideration

Implement sustainable practices to minimize waste, conserve water, and reduce energy consumption.

11. Employee Training

Train staff on propagation techniques, plant care, and safety protocols.

12. Research and Innovation

Stay updated with the latest horticultural research and technology to improve plant quality and management practices.

13. Customer Service

Provide guidance to customers on plant selection, care instructions, and landscaping advice.

Effective nursery management requires a combination of horticultural knowledge, attention to detail, and business skills to produce high-quality plants for various purposes, from landscaping to food production.

Fruits

1

Introduction to Fruit Crops

Fruit crops play a vital role in human nutrition, providing essential vitamins, minerals, and dietary fiber. They are not only delicious but also contribute to the overall health and well-being of individuals. Fruit crops are cultivated worldwide and come in a diverse range of flavors, shapes, sizes, and colors. These crops are not only consumed fresh but are also processed into various products such as juices, jams, jellies, and dried fruits.

Here are some key points to consider when introducing fruit crops:

Diversity of Fruit Crops

Fruit crops encompass a wide variety of plants, each with its unique characteristics. Some common fruit crops include apples, oranges, bananas, grapes, strawberries, peaches, pears, and watermelons. There are also lesser-known fruits like dragon fruit, persimmon, and pawpaw that offer unique flavors and nutritional benefits.

Nutritional Value

Fruits are rich sources of essential nutrients, including vitamins (such as vitamin C, vitamin A, and various B vitamins), minerals (potassium, magnesium), and dietary fiber. Different fruits offer distinct nutritional profiles, making it important to include a variety of fruits in one's diet.

Cultivation and Growing Conditions

Fruit crops have varying requirements for soil type, climate, and growing conditions. Some fruits thrive in temperate climates, while others are better suited for tropical or subtropical regions. Proper soil preparation, irrigation, pest management, and pruning are essential for successful fruit cultivation.

Propagation

Fruit crops can be propagated through various methods, including seeds, cuttings, grafting, and budding. Propagation methods vary depending on the type of fruit and the desired characteristics of the new plants.

Pests and Diseases

Fruit crops are susceptible to a range of pests and diseases that can affect both yield and fruit quality. Integrated pest management (IPM) strategies that combine biological, chemical, and cultural control methods are commonly employed to mitigate these issues.

Harvesting and Post-Harvest Handling

Timing of harvest is crucial to ensure optimal fruit quality. Different fruits have different indicators of ripeness, such as color, texture, and aroma. Proper post-harvest handling, including washing, sorting, and storage, helps prolong shelf life and maintain fruit freshness.

Economic and Cultural Importance

Fruit cultivation contributes significantly to the global agricultural economy and provides livelihoods to millions of farmers and laborers worldwide. Additionally, many cultures have deep-rooted traditions and celebrations associated with fruit harvesting.

Environmental Impact

Fruit crop cultivation, like any form of agriculture, has an environmental impact. Sustainable farming practices, such as agro forestry, organic farming, and responsible water management, are important to minimize negative effects on ecosystems.

In conclusion, fruit crops are not only a source of nutrition and income but also a reflection of biodiversity and cultural heritage. Their cultivation requires careful attention to a range of factors, from cultivation practices to post-harvest handling, to ensure the delivery of fresh, flavorful, and nutritious produce to consumers around the world.

2

Fruit Cultivation: Nurturing Orchards of Abundance

Tree fruit cultivation, a fundamental branch of horticulture, revolves around the art and science of growing fruit-bearing trees that yield a diverse range of delectable and nutritious produce. From orchards of apples and oranges to groves of cherries and peaches, tree fruit cultivation embodies a balance of skill, knowledge, and care to ensure optimal yields, quality, and sustainability.

Orchard Establishment and Management: Crafting Productive and Thriving Fruit Havens

Orchard establishment and management form the foundation of successful tree fruit cultivation, encompassing a spectrum of practices that determine the health, productivity, and longevity of fruit-bearing trees. From selecting suitable sites to orchestrating efficient cultivation techniques, this intricate process lays the groundwork for orchards that yield abundant and high-quality fruits.

- **Site Selection and Preparation:** Choosing the right site is paramount for orchard success. Factors such as soil type, drainage, sunlight exposure, and climatic conditions influence tree growth and fruit production. Site preparation involves clearing debris, tilling soil, and addressing drainage issues to create a hospitable environment for tree roots.
- **Selection of Tree Varieties:** Selecting appropriate tree varieties is a critical decision. Varieties should align with local climate and market demands. Disease-resistant cultivars are preferable to minimize the risk of pests and diseases.

- **Planting Techniques:** Proper planting techniques are essential to ensure healthy root establishment. Trees should be planted at the correct depth, with adequate spacing between them. Root ball structure, soil amendments, and mulching contribute to successful tree establishment.
- **Irrigation and Water Management:** Orchard trees require consistent and appropriate irrigation. Drip irrigation and other water-efficient methods optimize water use, preventing both over- and under-watering. Monitoring soil moisture and adjusting irrigation schedules based on tree needs and weather patterns are integral to water management.
- **Fertilization and Soil Health:** Maintaining soil fertility through balanced fertilization is crucial for robust tree growth and fruit production. Soil tests guide the application of appropriate nutrients. Organic matter incorporation and cover cropping enhance soil structure and health.
- **Pruning and Training:** Pruning shapes tree structure, encourages optimal light penetration, and stimulates fruit-bearing wood. Regular pruning removes dead or diseased branches and maintains open canopies. Training techniques dictate the shape of the tree, influencing fruit distribution and ease of management.
- **Pollination Enhancement:** Orchards reliant on pollination benefit from techniques that attract and support pollinators. Planting flowering cover crops, providing nesting sites, and minimizing pesticide use during bloom contribute to successful pollination.
- **Pest and Disease Management:** Effective Integrated Pest Management (IPM) strategies involve cultural practices, biological control, and judicious pesticide application. Regular monitoring of pests and diseases allows for timely interventions, reducing the need for chemical treatments.
- **Harvesting and Post-Harvest Handling:** Harvesting at the right stage of maturity ensures quality fruits. Gentle handling during harvesting minimizes bruising. Post-harvest practices such as sorting, grading, and cold storage maintain fruit quality until they reach consumers.
- **Sustainability and Innovation:** Orchard management is increasingly aligned with sustainability principles. Using cover crops, adopting organic practices, and implementing precision technologies enhance resource efficiency and minimize environmental impact.

Orchards are intricate ecosystems that demand holistic management approaches. Orchestrating proper establishment, attentive care, and innovative techniques results in orchards that thrive, bearing fruits that are not only delicious but also a testament to the dedication of horticulturists in nurturing nature's bounty.

Pruning and Training: Sculpting Healthy Trees for Optimal Yield and Structure

Pruning and training are integral practices in tree fruit cultivation, acting as both an art and a science to shape trees for maximum productivity, health, and longevity. These practices involve strategic removal of branches and guiding the growth of trees to create well-structured canopies that facilitate efficient light penetration, airflow, and fruit development.

Pruning Objectives:

- **Fruit Production:** Pruning removes excess branches and foliage, directing the tree's energy towards fruit production. Improved light exposure to inner branches ensures uniform fruit ripening and reduces the risk of diseases caused by poor air circulation.
- **Size Control:** Pruning helps manage tree size, preventing overcrowding and facilitating easier management tasks such as spraying, harvesting, and maintenance.
- **Disease Prevention:** Proper spacing and thinning of branches reduce the humidity within the canopy, minimizing conditions conducive to disease development. Removing diseased or dead branches prevents the spread of pathogens.
- **Shape and Aesthetics:** Pruning allows horticulturists to sculpt trees into desired shapes, optimizing sunlight exposure and enhancing the orchard's visual appeal.

Training Techniques for Optimal Fruit Tree Growth and Productivity

Training fruit trees involves shaping and directing their growth to maximize sunlight exposure, facilitate air circulation, and promote efficient fruit production. Proper training enhances orchard management, disease prevention, and ease of harvesting. Various training methods are used, tailored to specific fruit types and growth habits. Orchardists employ these techniques to create strong, healthy, and productive fruit trees that yield high-quality fruits and contribute to sustainable agriculture.

Central Leader Training

- **Applicable Trees:** Often used for apple and pear trees, as well as some stone fruits.
- **Method:** A single, dominant central trunk is encouraged, with lateral branches growing outward at different heights. This creates a pyramid or Christmas tree shape.
- **Advantages:** Allows for efficient light penetration, easy pruning, and uniform fruit distribution.

Open Center Training (Vase or Goblet Training):

- **Applicable Trees:** Suited for stone fruits like peaches, plums, and cherries.
- **Method:** Trees are trained to have an open center with a vase-like shape. The central leader is removed, and several main branches grow outwards.
- **Advantages:** Facilitates sunlight penetration, good air circulation, and accessibility for harvesting and maintenance.

Espalier Training:

- **Applicable Trees:** Suitable for apples, pears, and other trees with flexible branches.
- **Method:** Trees are trained to grow flat against a wall, fence, or trellis, with branches extending horizontally.
- **Advantages:** Saves space, increases sunlight exposure, and facilitates pest and disease management.

Palmette Training:

- **Applicable Trees:** Often used for apple and pear trees.
- **Method:** Trees have a flat, two-dimensional shape with multiple tiers of horizontal branches.
- **Advantages:** Maximizes fruit-bearing surface area, simplifies pruning, and allows for good light distribution.

Trellis Training:

- **Applicable Trees:** Applicable to grapes and some tree fruits.

- ✯ **Method:** Trees or vines are supported by a trellis system, guiding growth vertically or horizontally.
- ✯ **Advantages:** Facilitates efficient space utilization, improves light exposure, and eases maintenance.

Topiary Training:

- ✯ **Applicable Trees:** Often used for ornamental fruit trees.
- ✯ **Method:** Trees are pruned and trained into artistic shapes, such as spirals or animal forms.
- ✯ **Advantages:** Aesthetic appeal, creative expression, and potential for artistic orchard designs.

Considerations:

- ✯ **Pruning Techniques:** Regular pruning complements training methods, removing diseased, dead, or crowded branches.
- ✯ **Tree Health:** Proper training encourages even fruit distribution, reducing the risk of branch breakage due to excessive fruit weight.
- ✯ **Varietal Suitability:** Different fruit varieties respond differently to training methods; it's crucial to select techniques that suit their growth habits.

Effective training techniques guide fruit trees toward optimal growth patterns, ensuring efficient resource utilization, ease of management, and high fruit quality. By aligning training methods with specific fruit types and orchard goals, orchardists create healthy, productive trees that flourish within the context of sustainable and environmentally conscious orchard management.

Pruning Practices: Shaping Strong, Healthy, and Fruitful Fruit Trees

Pruning is a vital horticultural practice that plays a key role in shaping the growth, health, and productivity of fruit trees. Done with precision and purpose, pruning enhances sunlight exposure, air circulation, and disease prevention, resulting in well-structured trees that yield high-quality fruits.

Objective-Oriented Pruning:

- ✯ **Structural Pruning:** In the early years, focus on creating a strong framework by removing weak, crossing, or competing branches. This establishes a solid foundation for future growth.

- ✯ **Fruit Production Pruning:** As trees mature, prioritize shaping the canopy to allow sunlight penetration and even fruit distribution. Remove overcrowded branches and those that hinder air circulation.

Basic Pruning Techniques:

- ✯ **Thinning:** Removing entire branches back to the main trunk or a lateral branch. Thinning reduces overcrowding, stimulates new growth, and enhances fruit quality.
- ✯ **Heading Back:** Shortening branches by cutting back to a lateral bud. This encourages branching and helps control tree height.
- ✯ **Pinching:** Removing the tips of young, tender shoots. Pinching encourages lateral branching and a bushier growth habit.
- ✯ **Notching:** Creating a small cut above a bud to redirect growth away from the bud. Notching influences the direction of new growth.

Key Considerations:

- ✯ **Pruning Timing:** Prune during the dormant season (late winter to early spring) when the tree is less susceptible to stress and disease.
- ✯ **Pruning Severity:** Avoid excessive pruning, as it can stress the tree and impact fruit production.
- ✯ **Proper Tools:** Use sharp, clean pruning tools to ensure clean cuts that heal quickly.

Specialized Pruning Techniques:

- ✯ **Thinning Cuts:** Remove entire branches to improve light penetration, reduce congestion, and maintain overall tree health.
- ✯ **Heading Cuts:** Cut just above an outward-facing bud to promote balanced branching and discourage inward growth.
- ✯ **Renewal Pruning:** Remove old, unproductive wood to encourage new growth and rejuvenate the tree.

Training Young Trees:

- ✯ **Central Leader:** Train the main trunk to grow vertically, with lateral branches forming at various heights. This creates a strong, balanced structure.
- ✯ **Open Center:** Encourage an open, vase-like structure with several main branches radiating from the trunk. This allows sunlight to reach the center of the canopy.

Mature Tree Pruning:

- **Canopy Management:** Regularly thin the canopy to maintain an open structure that facilitates light penetration and air circulation.
- **Fruit Load Management:** In heavy fruiting years, thin excess fruit to prevent branches from breaking under the weight.

Pruning is a skill that merges science and art, guiding fruit trees toward their full potential. Through thoughtful selection of pruning techniques, orchardists balance growth, fruit production, and tree health. By nurturing trees with precision, orchardists create resilient, productive orchards that reflect the artistry of nature and human intervention working harmoniously together.

Pruning Practices: Customized Techniques for Different Horticultural Crops

Pruning techniques vary based on the specific growth habits, fruiting characteristics, and objectives of different horticultural crops. From trees to vines and shrubs, each type of plant demands tailored pruning to ensure optimal growth, yield, and overall health. Let's explore pruning practices with examples of common horticultural crops.

1. Fruit Trees:

- **Apple (*Malus domestica*):** Apple trees benefit from central leader training in their early years. Remove competing branches and maintain an open center for sunlight penetration. As they mature, thin out crowded branches and perform renewal pruning by removing older wood to encourage new growth.
- **Peach (*Prunus persica*):** Open center training suits peach trees. Remove the central leader and encourage outward growth of main branches. Regularly thin out fruit clusters to prevent overcrowding and maintain manageable fruit size.

2. Grapevines:

- **Grapes (*Vitis vinifera*):** Grapevines require specialized pruning to optimize fruit production. For spur-pruned varieties, prune to two-bud spurs along the main trunk. For cane-pruned varieties, leave one-year-old canes with 8 to 12 buds, removing older wood.

3. Berry Bushes:

- **Blueberries (*Vaccinium spp.*):** Remove low-lying branches and dead wood. In winter, perform cane renewal by cutting out the oldest canes at ground level to promote new growth.
- **Blackberries (*Rubus spp.*):** Prune trailing blackberries to maintain a manageable size and encourage new growth. Cut back canes that have fruited to the ground and tie up new canes for next year's fruiting.

4. Roses:

- **Hybrid Tea Roses (*Rosa* spp.):** Prune in late winter to remove dead wood and thin out crossing branches. Cut back to an outward-facing bud to encourage an open, vase-like shape.
- **Climbing Roses**: Remove any dead or weak wood and prune lateral branches to encourage new flowering shoots.

5. Shrubs:

- **Lilacs (*Syringa* spp.):** After blooming, remove spent flowers to prevent seed formation. Thin out older wood to encourage new growth from the base.
- **Ornamental Trees (e.g., *Japanese Maple, Acer palmatum*):** Prune to maintain a balanced canopy and remove any dead or diseased branches. Use selective thinning cuts to enhance the tree's shape and structure.

6. Ornamental Vines:

- **Clematis:** Group 1 clematis are pruned after blooming to remove dead or weak stems. Group 2 clematis are pruned to remove dead or damaged wood, leaving healthy buds to flower.
- **WisteriA.** In late winter, prune to remove any tangled or excessive growth. Cut back long stems to encourage the formation of flower buds.

Pruning practices are as diverse as the horticultural crops they shape. By understanding the unique growth habits and requirements of each type of plant, orchardists and gardeners can tailor their pruning techniques to promote healthy growth, optimal fruit production, and aesthetic beauty. Customized pruning ensures that each plant thrives in its own way, contributing to the overall health and productivity of the orchard or garden.

Pollination and Fruit Set: Nature's Partnership for Abundant Harvests

Pollination and fruit set are fundamental processes in tree fruit cultivation, orchestrating the delicate dance between plants and pollinators that culminates in the formation of delectable and nutritious fruits. These intertwined phenomena hold the key to successful fruit production, impacting crop yields, quality, and the sustainability of orchards.

Pollination and Its Importance:

- ✯ **The Role of Pollinators:** Pollinators, primarily bees but also butterflies, moths, and other insects, facilitate the transfer of pollen from male flower parts (anthers) to female flower parts (stigmas). This process is essential for fertilization, enabling the formation of seeds within the fruit.
- ✯ **Enhanced Fruit Quality:** Adequate pollination leads to optimal fruit set, resulting in larger, more uniform, and better-shaped fruits. Proper pollination also improves fruit color, sweetness, and overall market value.
- ✯ **Crop Yield:** A strong correlation exists between pollination efficiency and crop yield. Inadequate pollination can lead to poor fruit set, resulting in smaller harvests and economic losses.

Factors Influencing Pollination:

- ✯ **Flower Structure:** Flower structure plays a role in pollination success. Open flowers with accessible anthers and stigmas are more likely to attract pollinators.
- ✯ **Timing:** Proper synchronization between flower opening and pollinator activity is crucial. Blooms should coincide with the peak activity of pollinators in the orchard.
- ✯ **Weather Conditions:** Pollination can be affected by weather factors such as temperature, humidity, and wind speed. Rain and strong winds can impede pollinator activity.

Managed Pollination:

- ✯ **Hive Placement:** In commercial orchards, managed honeybee colonies are often brought in during bloom to ensure sufficient pollination. Hive placement is strategic to optimize pollinator coverage.

- **Supplemental Pollination:** In cases of inadequate natural pollination, supplemental pollination using techniques like hand pollination or the distribution of pollen via blowers can boost fruit set.

Fruit Set and Fruit Development:

- **Fruit Set:** After successful pollination, fertilized flowers undergo fruit set—the stage when ovaries start to develop into fruits. Proper pollination leads to higher fruit set rates.
- **Thinning:** In some cases, fruit thinning may be necessary to achieve optimal fruit size and quality. Thinning involves manually removing excess fruits to avoid overcrowding and allow remaining fruits to develop fully.

Pollination and fruit set are intricate ecological interactions that underscore the symbiotic relationship between plants and pollinators. Ensuring the health and availability of pollinators, optimizing flower conditions, and fostering an environment conducive to successful pollination contribute to abundant fruit yields and the continuation of this vital partnership between nature and agriculture.

Many tree fruit species rely on effective pollination for fruit set. Bees and other pollinators play an essential role in transferring pollen between flowers, leading to successful fertilization and fruit development. Orchards may require managed pollination to ensure optimal yields.

Pest and Disease Management: Safeguarding Orchard Health and Productivity

Pest and disease management in tree fruit cultivation is a dynamic and essential aspect that safeguards the health, quality, and yield of orchards. Orchestrating integrated strategies that balance preventive measures, biological controls, and judicious pesticide use is crucial for maintaining a thriving and sustainable fruit production system.

Integrated Pest Management (IPM):

- **Holistic Approach:** IPM is a comprehensive strategy that integrates various practices to manage pests and diseases. It emphasizes proactive measures over reactive treatments, fostering ecological balance within the orchard ecosystem.

- **Monitoring and Identification:** Regular monitoring of pest and disease populations allows growers to identify potential issues early. This facilitates informed decisions regarding the timing and necessity of interventions.
- **Cultural Practices**: Cultural practices, such as proper sanitation, orchard floor management, and maintaining tree vigor through proper fertilization and irrigation, create an environment less conducive to pests and diseases.

Biological Controls: Beneficial organisms, like predatory insects, parasitic wasps, and mites, can be introduced to regulate pest populations naturally. These biological controls minimize the reliance on chemical interventions.

Pest Management:

- **Insect Pests:** Insects like aphids, fruit flies, and codling moths can damage fruit quality and reduce yields. Monitoring, trapping, and using pheromone-based techniques can help manage their populations.

Disease Management:

- **Fungal Diseases:** Diseases like apple scab, brown rot, and powdery mildew can impact fruit quality. Cultural practices such as pruning to increase air circulation and fungicide application are employed for control.
- **Bacterial Diseases:** Bacterial pathogens can cause diseases like fire blight in apples and pears. Preventive measures like pruning affected branches and applying antibacterial agents can limit their spread.

Chemical Interventions:

- **Selective Pesticides:** When necessary, selective pesticides are used to target specific pests while minimizing harm to beneficial organisms and the environment.
- **Residue Management:** Adhering to recommended waiting periods between pesticide application and harvest ensures that fruits have minimal pesticide residues.
- **Resistance Management:** Rotating different classes of pesticides and using them judiciously reduces the likelihood of pests developing resistance.

Preventive Measures:

- ✯ **Plant Health:** Strong, healthy trees are less susceptible to pests and diseases. Ensuring proper nutrition, irrigation, and soil health boosts tree resilience.
- ✯ **Quarantine:** Preventing the introduction of new pests and diseases through rigorous quarantine measures for plant material, equipment, and personnel is crucial.

Education and Awareness:

- ✯ **Grower Training:** Educating growers about pest and disease identification, IPM strategies, and the responsible use of pesticides enhances their capacity for effective management.
- ✯ **Community Involvement:** Engaging communities in pest and disease awareness programs encourages vigilance and cooperation in managing orchard health.

Pest and disease management is a continuous and adaptive endeavor, guided by the principles of sustainability, ecosystem health, and effective resource management. Through a blend of science, innovation, and responsible practices, orchardists strive to strike a harmonious balance that ensures bountiful harvests while preserving the health of orchards and the environment.

Thinning and Fruit Development: Nurturing Quality and Yield Through Prudent Management

Thinning and fruit development are vital stages in tree fruit cultivation that directly influence the size, quality, and marketability of the harvested produce. These practices involve meticulous decision-making and careful interventions to optimize fruit growth, prevent over-crowding, and enhance overall orchard productivity.

Importance of Thinning:

- ✯ **Enhancing Fruit Size:** Thinning aims to ensure that the remaining fruits have enough space to grow to their optimal size. Overcrowded fruits are often smaller, misshapen, and of lower quality.
- ✯ **Reducing Biennial Bearing:** Thinning breaks the biennial bearing cycle, in which trees produce an excessive crop one year followed by a lower yield the next. By reducing the fruit load, trees are more likely to produce a consistent yield each year.

- **Minimizing Branch Breakage:** Heavy fruit loads can cause branches to break under the weight, leading to costly damage and reduced future yields.
- **Improving Color and Flavor:** Properly thinned fruits receive more sunlight and airflow, resulting in better coloration and sweeter flavors due to increased sugar content.

Thinning Techniques:

- **Hand Thinning:** Skilled workers manually remove excess fruits by hand. This method is precise and allows for selective thinning based on fruit size and position.
- **Chemical Thinning:** Chemicals are used to induce fruit drop by disrupting hormone balance. This method requires accurate application timing to avoid excessive thinning.
- **Mechanical Thinning:** Mechanical thinners, equipped with rotating brushes or pads, gently remove fruits. This method is efficient for large orchards but requires calibration to prevent over-thinning.

Factors Influencing Thinning:

- **Natural Fruit Drop:** Trees often undergo a natural thinning process, known as June drop, where a certain number of fruits naturally fall off. However, supplemental thinning may still be necessary.
- **Fruit Load:** The number of fruits per tree varies by species and cultivar. Proper thinning aims to balance fruit load for optimal growth.
- **Tree Vigor:** Vigorous trees may require more aggressive thinning to prevent over-crowding and maintain fruit quality.
- **Weather Conditions:** Cold or frosty weather during flowering can lead to poor fruit set and may require less thinning.

Fruit Development:

- **Nutrient Supply:** Adequate nutrient availability, particularly during the early stages of fruit development, is essential for optimal growth.
- **Water Management:** Consistent and appropriate irrigation is crucial to ensure continuous growth and prevent fruit drop due to water stress.
- **Protection from Pests and Diseases:** Fruit development requires vigilant pest and disease management to prevent damage that could lead to premature fruit drop.

3

Berry & Small Fruit Cultivation: Harnessing Nature's Jewels for Nutritious Delights

Berry and small fruit cultivation represent a vibrant and dynamic segment of horticulture, encompassing a diverse array of fruits that are celebrated for their vibrant flavors, nutritional benefits, and versatility. From the lush fields of strawberries to the trellised vines of grapes, cultivating these miniature marvels requires a unique blend of techniques and expertise to ensure optimal yields, quality, and sustainability.

Variety and Diversity

- ★ **Berries:** Berries include strawberries, blueberries, raspberries, and blackberries, each boasting distinctive flavors, colors, and growth habits. Their succulent nature and rich antioxidant content make them highly sought after in the culinary and health industries.
- ★ **Small Fruits:** Small fruits encompass a broader category, including grapes, currants, gooseberries, and kiwi. These fruits often grow in clusters or bunches, offering both fresh consumption and processing potential.

Here are some common types of berries and small fruits, along with cultivation details:

1. Strawberries:

- ★ **Cultivation:** Strawberries are usually grown as perennial plants. They can be cultivated in the ground, raised beds, containers, and hanging baskets. They require well-drained soil with good organic content.
- ★ **Propagation:** Strawberries can be propagated from runners (stolons) or by purchasing certified disease-free plants.
- ★ **Growing Conditions:** Strawberries prefer full sun and slightly acidic soil (pH 5.5 to 6.5). Adequate spacing and good air circulation help prevent diseases.
- ★ **Harvest:** Berries are harvested when fully ripe. Harvesting methods depend on the type of strawberry (June-bearing, everbearing, or day-neutral).

2. Blueberries:

- ★ **Cultivation:** Blueberries are woody shrubs that thrive in acidic soil with good drainage. They can be grown in the ground or containers.
- ★ **Growing Conditions:** Blueberries prefer full sun to partial shade and require consistent moisture.
- ★ **Pruning:** Regular pruning helps maintain plant health and encourage fruit production. Blueberries bear fruit on previous year's growth.
- ★ **Harvest:** Berries are typically harvested when they are fully ripe. They should come off the plant easily.

3. Raspberries:

- ★ **Cultivation:** Raspberries can be grown as erect or trailing canes. They need support systems like trellises for proper growth.
- ★ **Propagation:** Raspberries can be propagated from root cuttings, suckers, or tissue culture.
- ★ **Growing Conditions:** Raspberries require well-drained soil, full sun, and good air circulation to prevent diseases.
- ★ **Pruning:** Pruning is essential for maintaining plant health and encouraging fruit production. Prune based on the type (summer-bearing or fall-bearing).
- ★ **Harvest:** Berries are harvested when they are fully colored and easily detach from the plant.

4. Blackberries:

- **Cultivation:** Blackberries are trailing canes that can be grown with trellises for support.
- **Growing Conditions:** They prefer full sun, well-drained soil, and consistent moisture.
- **Pruning:** Pruning helps manage plant growth and encourage fruiting. Different varieties require different pruning methods.
- **Harvest:** Blackberries are harvested when fully ripe and easily detach from the plant.

5. Currants:

- **Cultivation:** Currants are grown as bushes and prefer well-drained soil with moderate fertility.
- **Growing Conditions:** They can tolerate partial shade but prefer full sun. Adequate spacing and air circulation are important for disease prevention.
- **Pruning:** Pruning is important to maintain plant shape, remove old wood, and encourage new growth.
- **Harvest:** Currants are harvested when the berries are fully colored and easy to pick.

6. Gooseberries:

- **Cultivation:** Gooseberries are grown as bushes and prefer well-drained soil.
- **Growing Conditions:** They can tolerate partial shade but prefer full sun. Prune to improve air circulation and reduce disease risk.
- **Harvest:** Gooseberries are harvested when fully ripe and have reached the desired size and color.

Sustainability and Innovation

- **Agroecological Practices:** Techniques like cover cropping, companion planting, and organic fertilization enhance soil health and minimize environmental impact.
- **Protected Cultivation:** Greenhouses and high tunnels extend the growing season, protect crops from adverse weather, and enhance control over environmental conditions.

Market Opportunities:

- **Fresh Consumption:** Berries and small fruits are favored for their fresh, flavorful, and nutrient-rich attributes.
- **Processed Products:** Processed products such as jams, jellies, juices, and wines offer diverse value-added opportunities.

Successful berry and small fruit cultivation requires attention to factors such as soil preparation, proper planting, irrigation, pest and disease management, and proper harvesting techniques. It's important to choose varieties that are well-suited to your climate and growing conditions. Regular care and maintenance, including fertilization and proper pruning, play a crucial role in achieving healthy and productive berry plants. Their cultivation is a testament to the delicate balance between nurturing healthy plants and harnessing the incredible flavors and nutritional benefits that these diminutive fruits offer to consumers, marking a delightful journey from farm to table.

4

Citrus & Tropical Fruit Cultivation: Orchards of Sun-Kissed Delights

Citrus and tropical fruit cultivation transport us to sun-drenched landscapes, where the vibrant flavors and exotic aromas of fruits like oranges, mangoes, and pineapples beckon. These fruits, emblematic of warm climates, thrive in carefully managed orchards where the convergence of expertise, climate, and dedication yields bountiful harvests of luscious treats.

- **Citrus Gems:** Oranges, lemons, limes, and grapefruits are among the citrus treasures, each offering unique taste profiles and culinary uses. Citrus fruits are rich in vitamin C and are integral components of diets worldwide.
- **Tropical Temptations:** Mangoes, pineapples, bananas, and papayas are tropical fruit staples. Their rich flavors, vivid colors, and nutritive value make them popular in both local markets and international trade.

Citrus crops can be classified into various groups based on factors like fruit type, use, and genetic traits. Here is a classification of citrus crops:

1. Sweet Oranges (*Citrus* × *sinensis*):

- Navel Oranges
- Valencia Oranges
- Blood Oranges
- Cara Cara Oranges

2. Mandarins and Tangerines (*Citrus reticulata*):

- Clementines
- Satsumas
- Tangerines

3. Grapefruits (*Citrus* × *paradisi*):

- White (or Yellow) Grapefruits
- Pink Grapefruits
- Ruby Red Grapefruits

4. Lemons (*Citrus limon*):

- Eureka Lemons
- Lisbon Lemons
- Meyer Lemons

5. Limes (*Citrus aurantiifolia* and *Citrus* × *latifolia*):

- Key Limes
- Persian Limes (Tahiti Limes)
- Kaffir Limes

6. Citron (*Citrus medica*):

- Etrog Citron
- Buddha's Hand Citron

7. Pomelos (*Citrus maxima*):

- Chandler Pomelos
- Honey Pomelos

8. Hybrid Citrus:

- Tangelos (tangerines x grapefruits)
- Tangors (tangerines x oranges)
- Ugli Fruit (cross between grapefruit, orange, and tangerine)

9. Specialty and Exotic Citrus:

- Kumquats
- Finger Limes (*Citrus australasica*)

- ★ Calamondins
- ★ Sudachi (*Citrus sudachi*)

Ornamental Citrus: Some citrus varieties are cultivated primarily for their ornamental value, such as the decorative and fragrant blooms. These may not be grown for fruit consumption.

Rootstocks: Citrus rootstocks are cultivated for their specific characteristics that influence tree size, adaptability to soil conditions, and resistance to diseases. Common rootstock varieties include sour orange, Carrizo citrange, and Swingle citrumelo.

Wild Citrus Relatives: These are wild species that are closely related to cultivated citrus and are often used in breeding programs for disease resistance and other desirable traits.

It's important to note that this classification is not exhaustive, as the citrus family includes numerous species, hybrids, and varieties that are cultivated and appreciated for their diverse flavors, aromas, and uses. Each type of citrus fruit brings its own unique qualities to the table, contributing to the rich tapestry of flavors found within the citrus crop family.

Cultivation of Citrus Crops

Citrus crops are among the most popular and economically significant fruit crops worldwide, appreciated for their juicy, tangy flavors and high nutritional value. This chapter provides a comprehensive overview of the cultivation practices involved in growing citrus crops.

Site Selection and Preparation

Prepare the planting site by clearing the land of weeds, debris, and competing vegetation. Incorporate organic matter into the soil to enhance its structure and fertility. Proper drainage is crucial to prevent water logging.

Climate and Soil Requirements

Citrus crops thrive in subtropical to tropical climates. They require warm temperatures for optimal growth and fruit development. Well-drained, sandy-loam soils with a pH range of 6 to 7.5 are ideal for citrus cultivation.

Propagation and Planting

Citrus crops can be propagated through various methods, including seeds, cuttings, and grafting. Budding and grafting onto disease-resistant rootstocks are commonly used techniques to ensure desirable fruit quality and disease resistance. Plant citrus trees during the warm season, avoiding extreme temperatures. Dig holes that are slightly larger than the rootball, and place the tree at the same depth it was in the nursery. Backfill the hole with soil and water thoroughly.

Orchard Management

- **Irrigation** Citrus trees need regular and consistent watering, especially during dry periods and fruit development. Drip irrigation is an efficient method that reduces water wastage and minimizes disease risks.
- **Fertilization** Apply balanced fertilizers that contain nitrogen, phosphorus, and potassium, along with micro nutrients. Adjust fertilizer applications based on soil tests and the tree's growth stage.
- **Pruning** Prune citrus trees to improve air circulation, light penetration, and overall tree health. Remove dead, diseased, or crossing branches. Pruning also helps manage tree size and shape.
- **Pest and Disease Management** Monitor for common pests like aphids, citrus leaf miners, and scales, as well as diseases such as citrus canker and citrus greening. Employ integrated pest management (IPM) strategies, including biological controls, cultural practices, and judicious use of pesticides.

Flowering and Fruit Development

Citrus trees typically flower in response to favorable conditions. Proper irrigation and nutrition play a role in promoting flowering. Weather conditions during flowering can impact fruit set. Fruits develop over several months. Thinning excess fruit clusters helps improve fruit size and quality. Monitor for pests and diseases that can affect fruit development.

Harvesting and Post-Harvest Management

Harvest citrus fruits when they reach appropriate maturity. Color, size, and taste are indicators of maturity. Fruits are usually hand-picked or harvested using specialized tools. Handle citrus fruits carefully to prevent bruising and damage. Sort fruits based on size and quality. Store them at cool temperatures to prolong shelf life. Waxing and surface treatments can enhance fruit appearance and storage life.

Cultivating citrus crops requires a combination of proper site selection, attentive orchard management, and effective pest and disease control. With the right care, these vibrant and flavorful fruits can thrive, contributing not only to agricultural economies but also to the enjoyment of people around the world.

Mango

Mango (*Mangifera indica*) is known as the "King of Fruits" due to its delicious taste, nutritional value, and cultural significance in many tropical and subtropical regions. This chapter explores the comprehensive process of cultivating mango trees, from choosing the right varieties to post-harvest management.

Selection of Varieties

Choosing Suitable Varieties Mango varieties vary in size, flavor, color, and adaptability to different climates. Selection should be based on local conditions, market demand, and disease resistance. Notable varieties include Alphonso, Kent, Haden, Tommy Atkins.

Site Selection and Preparation

Climate and Soil Requirements Mango trees thrive in tropical and subtropical climates with a dry period that facilitates flowering. They prefer well-drained soils with a pH ranging from 5.5 to 7.5. Soil should be rich in organic matter and free from water logging.

Site Preparation Clear the land of weeds and debris. Dig pits of about 1 meter in depth and diameter, incorporating organic matter and topsoil. Proper spacing between trees is essential for good air circulation and sunlight penetration.

Propagation and Planting

Mango trees can be propagated through seeds or vegetative methods like grafting, budding, or air-layering. Grafting onto root stocks is a common method as it imparts desirable characteristics and controls tree size. Plant during the rainy season to ensure initial establishment. Place the seedling or grafted tree in the center of the pit at the same level it was in the nursery. Fill the pit with soil, firm it around the tree, and create a water basin.

Orchard Management

Irrigation Established mango trees need regular watering, especially during dry spells. Drip irrigation is effective in minimizing water wastage and providing targeted hydration.

Fertilization Mango trees have high nutritional requirements. Apply a balanced fertilizer, rich in nitrogen, phosphorus, and potassium, based on soil tests. Split the annual dose into several applications throughout the year.

Pruning Prune mango trees to maintain an open canopy, allowing sunlight to reach inner branches. Remove dead, diseased, or crowded branches to improve air circulation and reduce disease pressure.

Pest and Disease Management Monitor for common pests like mango fruit fly and diseases like anthracnose. Implement integrated pest management (IPM) strategies, including cultural practices, biological controls, and judicious use of pesticides.

Flowering and Fruit Development

Mango trees typically flower once a year, often in response to dry and cooler conditions. Flower induction can be encouraged through controlled irrigation and temperature manipulation.

After successful pollination, mango fruits form and develop over several months. Thinning of excess fruit clusters can improve fruit size and quality. Protect developing fruits from pests and diseases.

Harvesting and Post-Harvest Management

Mangoes are generally harvested when mature but still firm. The color and aroma of the fruit serve as indicators of maturity. Fruits are harvested using special tools or by hand-picking.

Handle mangoes with care to avoid bruising and damage. Sort fruits based on size and quality. Store at cool temperatures to extend shelf life. Various treatments, such as hot water treatment, can be applied to control fruit flies.

Mango cultivation is a complex but rewarding endeavor. Successful production involves selecting appropriate varieties, preparing the site, nurturing the trees through proper management practices, and ensuring effective pest and disease control. By following these guidelines, farmers can enjoy bountiful harvests of this beloved tropical fruit while contributing to the rich agricultural tapestry of mango-growing regions worldwide.

Banana

Bananas are a popular and widely consumed fruit known for their energy-rich content, portability, and versatility in culinary applications. The classification of banana fruit can be based on various factors, including genetic traits, uses, and physical characteristics. Here's a classification of banana fruit:

Dessert Bananas (*Musa sapientum*):

- ★ **Cavendish Bananas:** The most common type of dessert banana, known for its sweet taste and creamy texture. Varieties include Grand Nain and Dwarf Cavendish.
- ★ **Gros Michel Bananas:** Once the dominant commercial banana variety, but susceptible to Panama disease. They have a rich, sweet flavor.

Cooking Bananas (*Musa paradisiaca*):

- ★ **Plantains:** Larger and starchier than dessert bananas, plantains are commonly cooked before consumption. They are used in various savory dishes and have a wide range of ripening stages.
- ★ **Green Bananas:** Unripe bananas used for cooking and often fried, boiled, or used in curries.

Specialty Bananas:

- ★ **Red Bananas (*Musa acuminata*):** Characterized by their reddish-purple skin and sweeter taste compared to yellow bananas.
- ★ **Blue Java Bananas (*Musa acuminata* 'Blue Java'):** Also known as "ice cream bananas" due to their texture and flavor resembling vanilla ice cream.
- ★ **Apple Bananas (*Musa acuminata* 'Apple'):** Small, short, and sweet bananas with a hint of apple flavor.
- ★ **Burro Bananas (*Musa acuminata* 'Burro'):** Short, stubby bananas with lemony flavor and creamy texture.

Plantain Hybrids:

- ★ **FHIA Hybrids:** Developed by the Honduran Foundation for Agricultural Research (FHIA), these hybrids combine characteristics of dessert and cooking bananas, aiming for disease resistance and improved yield.

Wild Banana Relatives:

- ★ **Fe'i Bananas:** A group of wild or semi-domesticated bananas found in the Pacific Islands, often consumed by local populations.

Ornamental Bananas:

- ✯ Some banana varieties are cultivated for their ornamental value rather than fruit production. Their large, lush foliage and fast growth make them popular in landscaping.

Hybrid Bananas:

- ✯ Various hybridization efforts have resulted in bananas with unique characteristics, including disease resistance, improved shelf life, and altered fruit traits.

Selection of Varieties

Choosing Suitable Varieties Banana cultivation involves selecting appropriate varieties based on intended use. Varieties can be broadly categorized into dessert bananas and cooking bananas (plantains). Popular dessert banana varieties include Cavendish and Gros Michel, while plantain varieties include Horn and French.

Site Selection and Preparation

Climate and Soil Requirements Bananas thrive in warm and humid climates with consistent temperatures and rainfall. They require a frost-free environment. Well-drained, fertile soils rich in organic matter are ideal for banana cultivation.

Site Preparation Prepare the site by clearing vegetation, removing debris, and plowing the land. Create mounds or ridges for improved drainage. Add organic matter to enhance soil fertility and structure.

Propagation and Planting

Bananas are typically propagated through suckers, which are shoots emerging from the base of mature plants. Select healthy suckers with well-developed roots for planting.

Plant suckers in a slanting position to prevent water accumulation in the growing tip. Planting depth should allow the growing point to be just above the soil surface. Space plants adequately for good air circulation.

Orchard Management

Irrigation Bananas require consistent moisture throughout the growing season. Provide regular irrigation, especially during dry periods. Drip irrigation and mulching can help retain soil moisture.

Fertilization Apply balanced fertilizers rich in potassium, nitrogen, and phosphorus. Regularly supplement with micronutrients like magnesium and boron. Divide fertilizer applications into multiple doses throughout the year.

Pest and Disease Management Monitor for pests like nematodes, weevils, and aphids, as well as diseases like Panama disease and Sigatoka leaf spot. Implement integrated pest management (IPM) strategies to minimize chemical use.

Flowering and Fruit Development

Flowering Banana plants produce a large inflorescence known as a "banana heart." Proper nutrition, water management, and climate conditions influence the development of the banana heart and subsequent fruiting.

Fruit Development Fruits develop in clusters known as "hands." Protect developing fruits from pests and diseases. Thinning excess hands can improve fruit size and quality.

Harvesting and Post-Harvest Management

Harvesting Harvest bananas when they reach maturity. For dessert bananas, they should be fully developed but still slightly green. For plantains, wait until the skin turns yellow with black spots.

Post-Harvest Handling Handle bananas with care to avoid bruising. Keep them in a cool, well-ventilated area. Post-harvest treatments like ethylene exposure can accelerate ripening.

Papaya

Papaya (*Carica papaya*) is a tropical fruit known for its vibrant flavor, succulent texture, and numerous health benefits. Cultivating papaya requires careful attention to climate, soil, and cultural practices to ensure a successful and productive harvest. Papaya cultivation is a significant agricultural activity in many parts of the world, including India.

Different regions have introduced and commercialized various papaya varieties to suit local climates, consumer preferences, and market demands. Here are some of the papaya varieties that have gained commercial prominence both in India and worldwide:

Red Lady (Taiwan Papaya):

- ✯ A popular papaya variety known for its sweet taste and appealing red-orange flesh.
- ✯ Grown extensively in India, especially in states like Andhra Pradesh, Karnataka, Tamil Nadu, and Maharashtra.

- ★ Red Lady papayas are prized for their uniform size, disease resistance, and high productivity.
- ★ Suited for both local consumption and export markets.

Coorg Honeydew:

- ★ Developed by the Indian Institute of Horticultural Research (IIHR) in Karnataka, India.
- ★ Known for its sweet flavor and unique honeydew-like aroma.
- ★ Coorg Honeydew papayas have gained popularity among Indian consumers due to their distinctive taste.

Pusa Delicious:

- ★ Developed by the Indian Agricultural Research Institute (IARI) in New Delhi, India.
- ★ This variety is notable for its large-sized, flavorful fruits with orange flesh.
- ★ Suited for both domestic consumption and commercial cultivation in India.

Surya

- ★ A papaya variety developed in India with a focus on disease resistance and high yield.
- ★ Known for its vibrant orange flesh and sweet taste.
- ★ Grown in various Indian states, contributing to the country's papaya production.

Solo Papayas (Various Varieties):

- ★ Solo papayas, including Sunrise and other cultivars, are grown extensively in India.
- ★ These varieties are preferred for their early maturation, compact size, and high yield.
- ★ Solo papayas cater to both local markets and export demands.

Maradol Papaya

- ★ A widely grown variety worldwide, including India.
- ★ Large-sized fruits with sweet flavor and distinctive aroma.
- ★ Grown in tropical and subtropical regions, including parts of India.

Red Maradol Papaya

- ✯ A variant of the Maradol papaya, known for its reddish-orange flesh and sweet taste.
- ✯ Grown in multiple countries, including India, to cater to diverse consumer preferences.

Tainung Papaya

- ✯ Originally developed in Taiwan and now grown in various parts of the world, including India.
- ✯ Tainung papayas are valued for their high yield, disease resistance, and good fruit quality.

Mexican Red Papaya

- ✯ A common papaya variety in both India and worldwide markets.
- ✯ Medium-sized fruits with sweet flavor and vibrant orange-red skin.
- ✯ Valued for their versatility in culinary applications and fruit processing.

Kapoho Solo:

- ✯ A dwarf papaya variety often cultivated in India for small gardens and home consumption.
- ✯ Compact growth habit and sweet-tasting fruits make it suitable for limited spaces.

These are just a few examples of the papaya varieties that have gained commercial prominence in India and around the world. The diversity of papaya varieties reflects the adaptability of the fruit to different growing conditions and the preferences of consumers in various regions.

Site Selection and Preparation:

- ✯ **Climate Requirements:** Papayas thrive in tropical and subtropical climates with consistent temperatures above 60°F (15°C). Frosts and cold snaps can be detrimental to papaya plants, so selecting a suitable site with a frost-free climate is essential.
- ✯ **Soil Conditions:** Well-draining, sandy loam soils with a pH range of 6.0 to 6.5 are ideal for papaya cultivation. Soil that retains excess water can lead to root rot, so proper drainage is crucial.
- ✯ **Site Preparation:** Prepare the planting site by clearing weeds, debris, and any potential sources of contamination. Amend the soil with organic matter to improve its structure and fertility.

Propagation and Planting:

- **Seed Selection and Germination:** Choose high-quality seeds from disease-free, mature fruits. Seeds can be germinated in seedling trays or directly in the ground. Provide consistent moisture and warmth during germination.
- **Transplanting Seedlings:** Transplant healthy seedlings into the prepared site when they have developed a few true leaves. Ensure proper spacing to allow ample room for growth.

Cultural Practices:

- **Watering:** Papayas require consistent moisture, especially during their early growth stages. Use a drip irrigation system to deliver water directly to the root zone and avoid wetting the foliage.
- **Fertilization:** Apply a balanced fertilizer that contains nitrogen, phosphorus, and potassium to promote healthy growth and fruit development. Regularly feed papaya plants every 4-6 weeks during the growing season.
- **Pruning:** Prune papaya plants to remove damaged or diseased leaves, promote air circulation, and encourage sturdy growth. Remove any male plants if you're aiming for all-female fruit-bearing trees.
- **Mulching:** Apply a layer of organic mulch around the base of papaya plants to conserve moisture, suppress weed growth, and maintain even soil temperatures.

Pest and Disease Management:

- **Common Pests:** Papayas are susceptible to pests like aphids, fruit flies, and mealybugs. Implement integrated pest management strategies, such as introducing natural predators and using organic insecticides.
- **Disease Prevention:** Diseases like papaya ring spot virus and powdery mildew can affect papaya plants. Choose disease-resistant varieties, practice crop rotation, and maintain good sanitation practices to prevent disease spread.

Harvesting and Post-Harvest Handling:

- **Determining Ripeness:** Papayas are ready for harvest when they change color from green to yellow or orange, depending on the variety. The fruit should yield slightly to gentle pressure.

- **Harvesting Techniques:** Cut the papaya from the tree using a sharp knife or pruning shears. Leave a small portion of the stem attached to the fruit.
- **Storage and Handling:** Handle papayas with care to prevent bruising. Store ripe papayas in a cool, dry place or in the refrigerator. Green papayas can be stored at room temperature to ripen.

Cultivating papaya is a rewarding endeavor that brings forth a tropical delight cherished by many. By adhering to proper site selection, soil preparation, and attentive cultural practices, growers can ensure healthy papaya plants that yield bountiful, juicy fruits. As they embark on this journey, orchardists embrace the vibrant world of papaya cultivation, nurturing trees that stand as a testament to nature's abundance and the fruits of human dedication.

Citrus and tropical fruit cultivation is an artful fusion of nature's bounty and human ingenuity. These fruits, emblematic of the tropics, beckon with their sensory allure and nutritional value. Orchards of citrus and tropical fruits stand as vibrant landscapes that intertwine tradition, innovation, and a profound appreciation for the sun-kissed delights that grace our plates and enrich our lives.

5

Grape and Vineyard Management: Crafting Elegance from Vines to Wine

Grape and vineyard management encapsulate the intricate journey of nurturing the grapevine, from its tender shoots to the transformation of its luscious fruit into the fine wines that grace tables around the world. This art and science require a delicate balance of agricultural prowess, terroir understanding, and wine making sensibility, resulting in vineyards that produce exceptional grapes for the creation of timeless libations.

Varieties and Terroir:

- **Grape Diversity:** From reds like Cabernet Sauvignon and Merlot to whites like Chardonnay and Sauvignon Blanc, grape varieties offer diverse flavor profiles for winemaking.
- **Terroir Influence:** Terroir, encompassing soil, climate, topography, and local factors, shapes the character of grapes, influencing their taste, aroma, and overall quality.

Site Selection and Preparation:

- **Sunlight Optimization:** Vineyards require ample sunlight exposure for photosynthesis and ripening. Sloped terrain aids drainage and sunlight penetration.
- **Soil Analysis:** Soil testing informs soil preparation and amendments to ensure optimal nutrient balance and water retention.

Vine Propagation and Planting:

- ✯ **Rootstock Selection:** Rootstock choice affects vine vigor, disease resistance, and adaptation to soil conditions.
- ✯ **Vine Spacing:** Proper spacing between vines allows for efficient light and airflow, minimizing disease pressure.

Training and Trellising:

- ✯ **Pruning Techniques:** Pruning establishes vine shape and regulates fruit-bearing shoots. Methods like cane and spur pruning impact yield and quality.
- ✯ **Trellis Systems:** Various trellis designs, such as vertical shoot positioning, enhance sunlight exposure and facilitate canopy management.

Pest and Disease Management:

- ✯ **Integrated Approaches:** IPM strategies combine cultural practices, natural predators, and targeted pesticide use to manage pests and diseases sustainably.
- ✯ **Vigilance:** Regular scouting and monitoring of vineyards enable early pest or disease detection, preventing potential damage.

Canopy Management:

- ✯ **Leaf Thinning:** Thinning leaves around grape clusters improves airflow and sunlight penetration, reducing the risk of diseases and enhancing fruit ripening.
- ✯ **Cluster Thinning:** Removing excess grape clusters directs the vine's energy towards producing fewer, higher-quality grapes.

Harvest Timing:

- ✯ **Brix Levels:** Grapes are harvested based on sugar content (Brix levels), acidity, and flavor development.
- ✯ **Hand vs. Mechanical Harvest:** Hand harvesting preserves grape integrity, while mechanical methods offer efficiency for large vineyards.

Wine Production:

- ✯ **Crushing and Fermentation:** Grapes are crushed to extract juice, which undergoes fermentation to produce wine.

- **Aging and Blending:** Wines may age in barrels, enhancing complexity. Blending different batches creates a desired flavor profile.

Sustainability and Innovation:

- **Organic and Biodynamic Practices:** Many vineyards adopt organic and biodynamic methods to minimize environmental impact and enhance soil health.
- **Precision Viticulture:** Technology like GPS mapping and remote sensing optimize vineyard management practices.

Grape and vineyard management weaves the story of agriculture, art, and culture. From carefully tended vines to the final pour of a well-crafted wine, this journey celebrates the marriage of human craftsmanship and nature's bounty, producing libations that embody the essence of the land and the hands that shape

6

Post-Harvest Handling of Fruits: Nurturing Freshness from Orchard to Table

The post-harvest handling of fruits is a critical phase that bridges the gap between orchard or field and the consumer's table. This intricate process encompasses a range of activities aimed at preserving the freshness, quality, and nutritional value of fruits, ensuring that they reach consumers in their prime state. From meticulous sorting to controlled storage, every step in post-harvest handling plays a pivotal role in delivering delectable and nutritious produce to markets worldwide.

Harvest Timing and Maturity: The Art of Picking at Perfection

Harvest timing and maturity are pivotal factors in the world of agriculture, particularly when it comes to ensuring the peak quality, flavor, and nutritional value of fruits. The delicate balance between waiting for fruits to reach their optimal stage of maturity and avoiding overripening or deterioration requires a keen understanding of each fruit's unique characteristics and the factors that influence their growth.

Optimal Harvest Stage:

- **Visual and Sensory Indicators:** Harvesting at the right time is guided by visual cues such as color, size, and uniformity. Fruits should exhibit the characteristic color of their variety.

- ✯ **Texture and Tactile Feedback:** Tactile assessments, like gently squeezing a fruit to feel for firmness, indicate ripeness. The ideal texture may vary depending on the fruit type.
- ✯ **Aroma and Smell:** The aroma emitted by fruits is a strong indicator of ripeness. A sweet, fragrant aroma often signifies that a fruit is ready to be harvested.

Flavor Development:

- ✯ **Sugar and Acid Balance:** Fruits reach their peak flavor when the balance between sugars and acids is optimal. Sugars contribute sweetness, while acids provide the necessary tang.
- ✯ **Aromatic Compounds:** Aromatic compounds contribute to the complex flavors of fruits. Waiting for these compounds to fully develop enhances taste.

Climate and Environmental Factors:

- ✯ **Sunlight Exposure:** Fruits exposed to adequate sunlight during their growth period tend to develop more intense flavors and vibrant colors.
- ✯ **Temperature:** Temperature influences the rate of fruit development. Warm temperatures generally hasten ripening, while cool temperatures can delay it.

Harvest Method and Use:

- ✯ **Immediate Consumption:** Fruits intended for immediate consumption are often harvested when they are fully ripe and at their peak flavor.
- ✯ **Transport and Storage**: Fruits destined for transportation or storage may be harvested slightly before peak ripeness to account for further ripening during transit.

Factors by Fruit Type:

- ✯ **Climacteric Fruits:** Fruits like bananas, tomatoes, and avocados continue to ripen after harvest due to the production of ethylene gas. Harvesting just before full ripeness is common to prevent overripening during transit.
- ✯ **Non-Climacteric Fruits:** Fruits like citrus and grapes do not ripen significantly after harvest. They are typically harvested when fully mature.

Uniformity and Consistency:

- **Commercial Considerations:** Commercial growers strive for uniformity in harvested fruits to meet market expectations and facilitate efficient handling.
- **Home Gardens:** Home gardeners may choose to harvest fruits at different stages to enjoy a prolonged harvest period and a variety of flavors.

Harvest timing and maturity epitomize the intersection of science and art in agriculture. The ability to discern the precise moment when a fruit reaches its peak of flavor, aroma, and nutritional value requires a deep connection with nature and an acute understanding of the intricate interplay of factors that contribute to each fruit's journey from farm to table.

Sorting and Grading:

- **Visual Inspection:** Fruits are visually assessed for size, color, shape, and overall appearance.
- **Quality Standards:** Grading involves categorizing fruits based on quality attributes, ensuring consistency in market presentation.

Importance of Sorting and Grading:

- **Quality Assurance:** Sorting and grading ensure that only fruits and vegetables meeting predetermined standards of size, color, and appearance are sent to market, maintaining consumer trust and satisfaction.
- **Marketability:** Consistently sized and aesthetically pleasing produce is more attractive to consumers, enhancing market demand and allowing for premium pricing.

Sorting Process:

- **Visual Inspection:** Trained workers visually inspect each piece of produce, assessing attributes such as color, size, shape, and blemishes.
- **Manual Sorting:** Some fruits, like berries and delicate items, are sorted by hand to prevent bruising or damage.
- **Mechanical Sorting:** Conveyor belts equipped with sensors, cameras, and automated systems detect imperfections and sort produce based on predetermined criteria.

Grading Criteria

- ★ **Size:** Produce is graded based on size, ensuring uniformity within each category. Larger or smaller pieces may fetch different prices or serve different market segments.
- ★ **Color and Appearance:** Vibrant color and a uniform appearance contribute to consumer preference. External blemishes, discoloration, or irregularities affect grading.
- ★ **Shape and Uniformity:** Uniformity in shape, such as uniform apple or potato shapes, is desired for packaging and presentation.
- ★ **Defects and Blemishes:** Produce is graded based on the presence of defects like cuts, bruises, scars, and insect damage.
- ★ **Numeric Grading:** Numeric codes or labels indicate quality levels, helping consumers make informed choices.
- ★ **Categories:** Grading categories, such as "Extra Fancy," "Fancy," and "Standard," provide clear distinctions in quality.

Sorting and Grading Facilities:

- ★ **Packinghouses:** Commercial packinghouses use advanced sorting and grading machinery to handle large volumes efficiently.
- ★ **Local Markets:** Small-scale operations at local markets may involve manual sorting and grading by producers themselves.

Value-Added Benefits:

- ★ **Waste Reduction:** Sorting and grading prevent substandard produce from reaching consumers, reducing food waste.
- ★ **Market Flexibility:** Different grades cater to diverse consumer preferences and market demands.

Consumer Confidence:

- ★ **Consistency:** Consistently graded produce builds consumer trust and loyalty.
- ★ **Expectation:** Consumers associate higher grades with better quality and are willing to pay a premium.

Sorting and grading are the art of selecting the finest gems from the harvest and presenting them to consumers in their most appealing form. The symphony of visual inspection, precision machinery, and human judgment ensures that each piece of produce shines, embodying the dedication to quality and the intricate nuances that make fresh fruits and vegetables a delight for the senses.

Cleaning and Sanitization: Purifying the Path from Harvest to Consumption

Cleaning and sanitization play a crucial role in the post-harvest handling of fruits and vegetables, ensuring that produce reaches consumers free from contaminants, dirt, and harmful microorganisms. This meticulous process involves removing surface impurities while maintaining the integrity of the produce, guaranteeing that the fresh, wholesome essence of each item remains intact and safe for consumption.

Importance of Cleaning and Sanitization:

- **Food Safety:** Cleaning and sanitization prevent the transmission of pathogens and reduce the risk of foodborne illnesses.
- **Quality Preservation:** Proper cleaning enhances the visual appeal and prolongs the shelf life of produce.

Cleaning Process:

- **Water Wash:** Rinsing produce with clean water removes visible dirt, debris, and some microorganisms.
- **Brushing or Scrubbing:** Soft brushes or cloth can be used to gently scrub the surface of items with thicker skins, like potatoes or melons.
- **Dip or Soak:** Some produce benefits from a brief dip or soak in water to loosen dirt or contaminants.

Sanitization Process:

- **Safe Sanitizers:** Approved food-safe sanitizers, such as chlorine-based solutions, hydrogen peroxide, or organic alternatives, are used to kill or reduce harmful microorganisms.
- **Concentration and Contact Time:** Proper concentration and sufficient contact time with the sanitizer are critical for effective sanitization.

Equipment and Tools:

- **Wash Basins or Tanks:** Stainless steel or food-grade plastic containers are used for washing and sanitizing.

- ✯ **Sprayers or Mist Systems:** These systems ensure even distribution of sanitizing solutions.

Best Practices:

- ✯ **Preventing Cross-Contamination:** Separate equipment for washing and sanitizing prevents cross-contamination from potential pathogens.
- ✯ **Avoiding Overcrowding:** Adequate space ensures that produce is evenly cleaned and sanitized.

Organic Produce:

- ✯ **Organic Sanitizers:** Organic producers use approved organic sanitizers to maintain the integrity of their products.
- ✯ **No Residue:** Organic sanitizers leave minimal or no residue on the produce.

Commercial vs. Home Settings:

- ✯ **Commercial Scale:** Packinghouses and processing facilities use automated washing and sanitizing systems to handle larger volumes.
- ✯ **Home Use:** Consumers can follow proper washing and sanitization practices at home to ensure food safety.

Cleaning and sanitization serve as sentinels of food safety, guarding the journey of fresh produce from farm to fork. This essential process, whether carried out in commercial settings or homes, assures consumers that the wholesome goodness of fruits and vegetables remains untainted by harmful contaminants, contributing to the enjoyment of healthy and nourishing meals.

Packaging

Packaging: Safeguarding Freshness and Preserving Perfection

Packaging is a pivotal stage in the post-harvest journey of fruits and vegetables, encapsulating their vibrant essence, ensuring protection, and facilitating their safe transport from fields to markets, and ultimately, to consumers' tables. This intricate process involves selecting appropriate materials, designing efficient formats, and employing advanced technologies that align with the unique characteristics of each produce item, while also fulfilling the needs of modern logistics and environmental considerations.

Importance of Packaging:

- **Protection:** Packaging shields produce from physical damage, bruising, and exposure to contaminants during transit and handling.
- **Preservation:** Proper packaging slows down the natural deterioration of produce, extending shelf life and maintaining freshness.

Packaging Materials:

- **Plastic:** Common for its versatility, plastic packaging can range from bags and clamshells to vacuum-sealed pouches.
- **Cardboard:** Sturdy cardboard boxes and crates are suitable for bulk quantities and provide protection during transportation.
- **Mesh Bags:** Mesh bags allow airflow, reducing condensation and preventing spoilage.
- **Biodegradable and Sustainable Materials:** Eco-friendly options, such as biodegradable plastics or plant-based materials, align with environmental consciousness.

Packaging Formats:

- **Bulk Packaging:** Used for larger quantities, bulk packaging is often seen in crates, boxes, and large containers.
- **Retail Packaging:** Designed for consumers, retail packaging includes trays, clamshells, and pre-portioned bags.

Packaging Considerations:

- **Ventilation:** Proper ventilation prevents moisture buildup, reducing the risk of mold and decay.
- **Protection from Light:** Certain packaging materials prevent light exposure, which can degrade the nutritional content of some produce.
- **Stack ability:** Efficient packaging allows for stacking without causing damage to lower layers.

Advanced Technologies:

- **Modified Atmosphere Packaging (MAP):** Modified atmosphere packaging adjusts the levels of oxygen, carbon dioxide, and humidity to slow down spoilage.
- **Smart Packaging:** Incorporating sensors that monitor temperature, humidity, and freshness levels helps track the condition of produce.

Labeling and Branding:

- ★ **Nutritional Information:** Labeling provides consumers with nutritional details, helping them make informed choices.
- ★ **Traceability:** Barcodes and QR codes enable tracking the origin and journey of produce, aiding recalls and transparency.

Environmental Considerations:

- ★ **Reducing Waste:** Eco-friendly packaging materials and designs aim to minimize waste and environmental impact.
- ★ **Recyclability:** Choosing recyclable materials aligns with sustainable practices.

Consumer Preferences:

- ★ **Convenience:** Packaging that allows for easy portioning and handling meets consumers' busy lifestyles.
- ★ **Visual Appeal:** Appealing packaging design enhances the perceived value of the produce.

Packaging is an art that weaves practicality, protection, and presentation into a cohesive whole. It envelops fruits and vegetables with care, ensuring that their vibrant beauty, flavors, and nutritional benefits remain intact, delighting consumers and extending the journey of nature's bounty far beyond the fields where they were grown.

Cold Chain Management

Cold Chain Management: The Art of Preserving Freshness in Transit

Cold chain management is a sophisticated and essential process that revolves around maintaining controlled temperatures throughout the entire journey of perishable goods, such as fruits and vegetables, from the point of harvest to the consumer's table. This intricate dance of temperature control, logistics, and technology ensures that the freshness, quality, and nutritional value of produce are preserved, allowing consumers to enjoy produce that is as vibrant as the day it was harvested.

Importance of Cold Chain Management:

- ★ **Preservation of Quality:** Cold chain management prevents spoilage, decay, and microbial growth, maintaining the freshness and taste of produce.

- **Safety Assurance:** Controlled temperatures inhibit the growth of pathogens, reducing the risk of foodborne illnesses.

Key Elements of Cold Chain Management:

- **Pre-Cooling:** Rapid cooling of harvested produce immediately after harvest prevents heat-related damage and extends shelf life.
- **Storage and Transport:** Refrigerated storage facilities and temperature-controlled transport vehicles maintain consistent cold temperatures.
- **Monitoring and Control:** Advanced sensors, data loggers, and remote monitoring systems ensure temperature accuracy and allow real-time adjustments.

Benefits of Cold Chain Management:

- **Extended Shelf Life:** Maintaining optimal temperatures slows down deterioration, extending the availability of fresh produce.
- **Global Trade:** Cold chain management enables the export and import of perishables across long distances.

Technological Innovations:

- **IoT and Sensors:** Internet of Things (IoT) technology and sensors provide real-time data on temperature, humidity, and conditions.
- **Blockchain:** Blockchain technology ensures transparency and traceability, aiding in tracking the entire cold chain journey.

Consumer Assurance:

- **Quality Expectation:** Consumers have higher expectations of fresh produce quality, making cold chain management essential.
- **Nutritional Value:** Maintaining low temperatures preserves the nutritional content of fruits and vegetables.

Cold chain management is an orchestration of science, technology, and logistics that ensures that the journey of fresh produce remains an unbroken sequence of controlled temperatures. It guarantees that the beauty, flavor, and nutritional benefits of fruits and vegetables reach consumers' tables, fulfilling the promise of fresh and wholesome goodness, regardless of the distance they have traveled.

Cold Chain Challenges:

- ★ **Temperature Fluctuations:** Even minor temperature fluctuations can impact produce quality, emphasizing the need for precision.
- ★ **Power Outages:** Backup power sources and contingency plans are vital to prevent disruptions in the cold chain.

Harvest and Handling:

- ★ **Harvest Timing:** Picking produce at the right stage of maturity and immediately cooling it maintains freshness.
- ★ **Proper Packing:** Adequate insulation and temperature control materials keep produce at desired temperatures.

Transport and Logistics:

- ★ **Refrigerated Vehicles:** Refrigerated trucks, ships, and planes are equipped with cooling systems to ensure consistent temperatures during transit.
- ★ **Temperature Maintenance:** Effective insulation and ventilation prevent temperature fluctuations caused by external factors.

Retail and Storage:

- ★ **Display and Storage:** Refrigerated displays and storage units at retail outlets maintain produce quality for consumers.
- ★ **Distribution Centers:** Temperature-controlled distribution centers act as hubs to consolidate and redistribute produce efficiently.

Quality Control

Quality Control: Ensuring Excellence in Fresh Produce

Quality control is a fundamental pillar in the post-harvest handling of fruits and vegetables, encompassing a series of systematic processes aimed at assessing, maintaining, and enhancing the overall quality, safety, and marketability of fresh produce. By employing a combination of visual assessments, scientific measurements, and adherence to established standards, quality control ensures that consumers receive produce that meets or exceeds their expectations in terms of appearance, taste, and safety.

Importance of Quality Control:

- ✯ Consumer Satisfaction: Quality control guarantees that consumers receive produce that meets their expectations, fostering trust and loyalty.
- ✯ Marketability: High-quality produce commands higher prices and creates a positive reputation for producers and retailers.

Quality Control Processes:

- ✯ **Visual Inspection:** Trained personnel assess produce for attributes like color, size, shape, and overall appearance.
- ✯ **Grading and Sorting:** Produce is categorized into different quality grades based on predetermined criteria.
- ✯ **Sensory Evaluation:** Taste panels evaluate the flavor, aroma, and texture of produce to ensure it meets desired standards.

Nutritional and Chemical Analysis:

- ✯ **Nutrient Content:** Laboratory analysis assesses the nutritional content of produce, ensuring its value for consumers.
- ✯ **Pesticide Residues:** Testing for pesticide residues ensures produce is within safe limits and complies with regulations.

Standard Operating Procedures (SOPs):

- ✯ **Uniformity:** SOPs ensure consistent grading and handling practices, preventing discrepancies in quality.
- ✯ **Traceability:** SOPs aid in tracking the origin and handling processes of produce, assisting in recalls if necessary.

Quality Parameters:

- ✯ **Physical Appearance:** Color, size, shape, and freedom from defects contribute to the visual appeal of produce.
- ✯ **Taste and Flavor:** Sensory attributes like sweetness, acidity, and aroma determine the flavor profile.
- ✯ **Texture:** Assessing the firmness, crispness, and juiciness of produce adds to its overall quality.

Quality Control Techniques:

- ✯ **Random Sampling:** Taking representative samples from batches for analysis ensures accuracy.

- ★ **Non-Destructive Testing:** Technologies like near-infrared spectroscopy and ultrasound assess internal quality without damaging the produce.

Quality Assurance:

- ★ **Regulatory Compliance:** Quality control ensures compliance with safety and quality regulations.
- ★ **Consistency:** Implementing effective quality control processes maintains consistent produce quality over time.

Consumer Confidence:

- ★ **Transparency:** Clear labeling and accurate representation of quality attributes build consumer trust.
- ★ **Safety:** Quality control measures reduce the risk of contaminants or pathogens in produce.

Technological Advancements:

- ★ **Machine Vision:** Advanced cameras and image processing systems assist in automated visual inspection.
- ★ **Blockchain:** Blockchain technology provides transparency and traceability in the supply chain, enhancing consumer confidence.

Quality control is a harmonious blend of science, precision, and consumer-centricity that ensures the journey of fresh produce is marked by excellence at every step. From the moment produce is harvested to its final presentation in markets and stores, quality control guarantees that the essence of nature's bounty remains untarnished, delivering satisfaction and nourishment to consumers around the world.

Value-Added Processing

Value-Added Processing: Elevating Fresh Produce to Culinary Delights

Value-added processing is a creative and innovative approach in the world of agriculture, transforming raw fruits and vegetables into an array of convenient, flavorful, and market-ready products. This intricate process involves various techniques, from simple washing and cutting to more complex processes like canning, drying, and juicing. By enhancing the convenience, flavor, and shelf life of fresh produce, value-added processing meets the evolving demands of consumers while reducing food waste.

Importance of Value-Added Processing:

- **Diversified Products:** Value-added processing offers a range of products, catering to diverse consumer preferences and culinary needs.
- **Minimizing Waste:** Utilizing imperfect or surplus produce for value-added products reduces food waste and loss.

Value-Added Techniques:

- **Washing and Cutting:** Pre-cut and washed fruits and vegetables offer convenience for busy consumers.
- **Drying:** Dehydrating fruits and vegetables preserves flavors and extends shelf life.
- **Canning and Bottling:** Sealing produce in cans or bottles using heat sterilization maintains freshness for extended periods.
- **Freezing:** Flash-freezing preserves nutritional content and texture, allowing year-round availability.
- **Juicing:** Extracting juices from fruits and vegetables offers a convenient and nutritious beverage option.
- **Fermentation:** Fermenting produce creates products like pickles, sauerkraut, and kimchi with enhanced flavors and health benefits.

Market Opportunities:

- **Convenience Foods**: Pre-cut, pre-packaged, and ready-to-eat products cater to busy lifestyles.
- **Health and Wellness:** Nutrient-rich juices, dried fruits, and fermented products align with health-conscious consumer trends.

Food Safety and Quality:

- **Hygiene:** Value-added processing must adhere to strict hygiene and safety protocols to prevent contamination.
- **Preservation Methods**: Techniques like canning and drying involve preservation methods that extend shelf life.

Flavor Enhancement:

- **Flavor Combinations:** Blending different fruits and vegetables creates unique and enticing flavor profiles.
- **Culinary Creativity:** Value-added processing allows for the incorporation of herbs, spices, and other ingredients to enhance taste.

Community and Sustainability:

- ✯ **Local Markets:** Value-added products support local producers and encourage community engagement.
- ✯ **Reducing Waste:** Utilizing imperfect or surplus produce reduces the environmental impact of food waste.

Consumer Preferences:

- ✯ **Convenience:** Ready-to-eat, pre-packaged products align with consumers' time-pressed lifestyles.
- ✯ **Dietary Needs:** Value-added products can be tailored to meet dietary preferences, including vegan, gluten-free, and organic.

Innovation and Culinary Trends:

- ✯ **Plant-Based Products:** Value-added processing contributes to the growing availability of plant-based alternatives.
- ✯ **Global Flavors:** Incorporating diverse cultural flavors in value-added products offers a taste of culinary diversity.

Value-added processing is a culinary alchemy that transforms raw produce into a spectrum of delightful products, from on-the-go snacks to gourmet treats. With each jar of pickles, packet of dried fruits, or bottle of fresh juice, value-added processing captures the essence of nature's bounty, presenting it in a form that resonates with modern lifestyles, culinary exploration, and sustainability ideals.

Traceability and Documentation:

- ✯ **Batch Tracking:** Proper labeling and documentation enable tracking fruits back to their origin, aiding recalls and quality assurance.
- ✯ **Regulatory Compliance:** Adherence to food safety regulations ensures that fruits meet established standards.

Market Distribution:

- ✯ **Timely Distribution:** Efficient logistics ensure that fruits reach retailers and consumers promptly.
- ✯ **Retail Displays:** Proper handling at retail outlets, including temperature control and frequent restocking, maintains fruit quality.

Consumer Education:

- **Handling Guidelines:** Providing consumers with proper storage and handling instructions at the point of purchase prolongs fruit freshness at home.
- **Minimizing Food Waste:** Educating consumers about using imperfect but edible fruits helps reduce food waste.

The post-harvest handling of fruits is a harmonious choreography that blends science, technology, and care. The orchestration of each step, from harvest to table, is essential to ensure that the vibrant colors, succulent flavors, and vital nutrients of fruits are preserved and appreciated by consumers worldwide.

7

Fruit Processing & Preservation: Harnessing Nature's Bounty for Longevity & Delight

Fruit processing and preservation is a culinary art and science that extends the lifespan of fruits while capturing their vibrant flavors and nutritional benefits. This intricate process involves transforming raw fruits into a variety of products, from jams and jellies to canned fruits and frozen treats. By preserving the essence of fruits through different techniques, producers ensure that the delight of these natural treasures can be enjoyed year-round.

Importance of Fruit Processing and Preservation

Fruit processing and preservation stand as crucial pillars in ensuring a consistent supply of fresh and nutritious produce for consumers throughout the year. This practice holds paramount importance in addressing food security and availability. By extending the shelf life of fruits beyond their natural harvest season, processing and preservation techniques bridge gaps between periods of abundance and scarcity, safeguarding against shortages that might arise due to factors like adverse weather conditions or disruptions in supply chains. This continuous availability of fruits contributes to a more balanced and diverse diet, promoting overall health and well-being.

One of the most compelling reasons for fruit processing and preservation is the reduction of food waste. In an era where food waste has emerged as a global concern, these practices provide an effective solution. Surplus or imperfect fruits that might be overlooked for fresh consumption can be transformed into valuable products. This not only curtails waste at the farm level but also minimizes losses along the supply chain. Through processing, these fruits find renewed purpose, ensuring that resources expended in their cultivation do not go to waste and that more people have access to nourishing food.

Beyond the realms of reducing waste, fruit processing and preservation contribute significantly to economic growth and the expansion of markets. These practices add value to raw produce, creating a diverse array of products that command higher market prices. This added value benefits farmers, processors, and retailers alike, bolstering local economies and creating employment opportunities. Moreover, the diversity of processed fruit products meets the evolving demands of consumers, promoting culinary creativity and allowing for the development of innovative offerings that cater to different tastes and preferences.

Sustainability also finds a home in the realm of fruit processing and preservation. By utilizing surplus or less visually appealing fruits, these practices minimize the environmental impact of discarded produce. Additionally, focusing on local and seasonal produce for preservation encourages a connection to regional harvests and promotes sustainable agricultural practices. This, in turn, supports local farmers and reduces the carbon footprint associated with transporting out-of-season fruits from distant locations.

In essence, the significance of fruit processing and preservation goes beyond the confines of extending shelf life or creating diverse products. It encompasses the realms of food security, waste reduction, economic prosperity, sustainability, and culinary innovation. By cherishing the art and science of these practices, we uphold the legacy of nature's bounty while addressing the challenges of a changing world.

- **Seasonal Extension:** Preservation allows consumers to enjoy fruits even when they are out of season.
- **Reducing Food Waste:** Processing surplus or imperfect fruits prevents waste and loss.

Principles of Preservation and Processing Techniques: Ensuring Longevity and Delight

The principles of preservation and processing techniques form the foundation of transforming perishable fruits into enduring delights that grace our tables throughout the year. Guided by scientific knowledge and culinary ingenuity, these principles ensure that the vibrant flavors, nutritional benefits, and aesthetic appeal of fruits are preserved while enhancing their versatility and convenience.

Principles of Preservation:

- **Control of Microorganisms:** Microbial growth is inhibited by techniques like heat treatment, which destroys pathogens and spoilage microorganisms.
- **Reduction of Enzyme Activity:** Enzymes responsible for ripening and spoilage are deactivated through processes like blanching or pasteurization.
- **Moisture Removal:** Dehydration reduces water content, preventing microbial growth and enzymatic activity.
- **Modification of Atmospheric Conditions:** Techniques like modified atmosphere packaging (MAP) alter the composition of gases around the produce, slowing down deterioration.

Processing Techniques:

- **Canning:** Fruits are sealed in airtight containers and heat-processed to destroy microorganisms, ensuring long shelf life without refrigeration.
- **Freezing:** Rapid freezing halts enzymatic activity and microbial growth, preserving the texture and nutritional content of fruits.
- **Drying:** Dehydration removes water, preventing spoilage, and making dried fruits lightweight and shelf-stable.
- **Jam and Jelly Making:** Cooking fruits with sugar and sometimes pectin creates thick spreads that inhibit microbial growth and enhance flavor.
- **Fruit Juices:** Extracting juices through mechanical or enzymatic methods captures the essence of fruits in liquid form.
- **Fermentation:** Controlled microbial fermentation enhances flavors, extends shelf life, and creates products like pickles and fermented fruit beverages.

Processing Techniques:

- **Canning:** Fruits are sealed in cans with sugar syrup or fruit juice, then heat-processed to extend shelf life.
- **Freezing:** Flash-freezing preserves the natural texture and nutritional content of fruits.

- ✯ **Drying:** Dehydrating fruits removes moisture, preventing spoilage and concentrating flavors.
- ✯ **Jam and Jelly Making:** Cooking fruits with sugar and pectin creates spreads that highlight natural sweetness.
- ✯ **Fruit Juices:** Extracting juices captures the essence of fruits in liquid form.
- ✯ **Fruit Leather:** Pureed fruits are dried into flexible sheets, creating a convenient and nutritious snack.

Flavor Enhancement:

- ✯ **Blending and Pairing:** Mixing different fruits or pairing them with herbs and spices creates unique flavors.
- ✯ **Concentration:** Drying or reduction techniques intensify flavors, creating more robust taste profiles.

Nutritional Retention:

- ✯ **Vitamins and Minerals:** Proper preservation methods maintain the nutritional integrity of fruits.
- ✯ **Antioxidants:** Some preservation methods, like drying, can even enhance antioxidant content.

Science and Technology:

- ✯ **Heat Treatment:** Techniques like blanching and pasteurization destroy harmful microorganisms and enzymes, ensuring safety and extending shelf life.
- ✯ **Dehydration:** By reducing water content, dehydration inhibits the growth of microorganisms, making fruits suitable for long-term storage.

Aesthetic Appeal:

- ✯ **Color Retention:** Preservation techniques aim to retain the natural color of fruits, ensuring visual appeal.
- ✯ **Texture Maintenance:** Proper processing ensures that the texture of fruits is preserved, preventing mushiness or degradation.

Nutritional Integrity:

- ✯ **Minimizing Nutrient Loss:** Techniques like freezing and minimal processing help retain the nutritional value of fruits.

- **Antioxidant Retention:** Some processing methods enhance antioxidant content, contributing to the health benefits of preserved fruits.

Culinary Creativity:

- **Ingredient Versatility:** Processed fruits serve as convenient ingredients for culinary experimentation, enhancing both sweet and savory dishes.
- **Flavor Enhancement:** Preserving fruits in different forms opens avenues for pairing with various herbs, spices, and complementary ingredients.

Economic and Social Impact:

- **Market Expansion:** Processed fruit products broaden the market, providing economic opportunities for growers, processors, and retailers.
- **Reducing Food Waste:** Utilizing surplus and less visually appealing fruits curtails waste and addresses food security challenges.

The principles of preservation and the diverse techniques employed in processing fruits reflect a harmonious blend of scientific precision and culinary artistry. These principles guide the transformation of nature's bounty into a spectrum of products that cater to diverse tastes, nutritional needs, and lifestyle demands. By embracing these principles, we celebrate the harmony between nature and human innovation, ensuring that the flavors and benefits of fruits remain accessible and delightful year-round.

Market Opportunities:

- **Convenience Foods:** Canned, frozen, and dried fruits offer convenience for cooking and snacking.
- **Bakery and Confectionery:** Processed fruits are used in a wide range of baked goods and confections.

Sustainability:

- **Reducing Waste:** Processing and preserving surplus fruits reduce the environmental impact of food waste.
- **Local and Seasonal:** Using locally sourced fruits promotes sustainability and supports local growers.

Innovation and Creativity:

- **Artisanal Products:** Craft jams, jellies, and pickled fruits cater to consumers seeking unique flavors.
- **Fusion Flavors:** Combining fruits with exotic spices or herbs creates innovative taste experiences.

Food Safety:

- ★ **Hygiene and Quality:** Strict hygiene practices prevent contamination during processing.
- ★ **Preservation Methods:** Proper techniques like sterilization and sealing prevent spoilage.

Consumer Preferences:

- ★ **Health and Wellness:** Natural, minimally processed fruits appeal to health-conscious consumers.
- ★ **Variety:** Processed fruits offer an array of options for different taste preferences and culinary uses.

Fruit processing and preservation is a symphony that harmonizes the beauty of nature with human ingenuity. Through this process, fruits transcend their ephemeral nature, becoming timeless delights that grace the breakfast table, adorn desserts, and inspire culinary creations. From jams that capture the essence of summer to frozen fruits that bring a burst of flavor to winter, the art of preservation celebrates the richness of the harvest and the magic of transforming it into enduring pleasures.

8

Integrated Pest Management in Fruit Orchards:

A Holistic Approach to Orchard Health

Integrated Pest Management (IPM) is a comprehensive and sustainable approach to managing pests and diseases in fruit orchards. Rooted in ecological principles, IPM seeks to minimize the use of synthetic pesticides while maximizing the health of the orchard ecosystem. By integrating various pest management strategies, IPM enhances long-term productivity, reduces environmental impact, and ensures the production of high-quality fruits.

Principles of IPM:

- **Monitoring and Identification:** Regular monitoring of pests, diseases, and beneficial organisms is crucial to understand the dynamics of the orchard ecosystem and identify potential problems.
- **Prevention:** Preventive measures, such as using disease-resistant plant varieties and maintaining proper orchard hygiene, reduce the initial establishment of pests.
- **Cultural Practices:** Implementing practices like proper pruning, spacing, and irrigation management enhances orchard health and reduces pest susceptibility.
- **Biological Control:** Encouraging the presence of natural enemies like predators and parasitoids helps control pest populations.
- **Chemical Control as a Last Resort:** Synthetic pesticides are used judiciously and only when non-chemical methods are insufficient. Selective and targeted applications minimize harm to beneficial organisms.

IPM Strategies in Fruit Orchards:

- **Biological Control:** Introducing or conserving natural predators and parasites of pests to regulate their populations.
- **Cultural Practices:** Proper sanitation, balanced nutrition, and orchard layout that promote good airflow can prevent pest infestations.
- **Chemical Control:** Carefully chosen pesticides, preferably biopesticides or those with minimal environmental impact, are used only when necessary and according to IPM guidelines.
- **Trap Crops:** Planting specific crops that attract pests away from main fruit crops can reduce direct damage.
- **Beneficial Insects:** Introducing beneficial insects like ladybugs or lacewings can control aphids and other pests.

Benefits of IPM:

- **Environmental Sustainability:** Reduced pesticide use minimizes harm to non-target organisms, soil health, and water quality.
- **Economic Savings:** Integrated approaches may lead to lower pesticide costs and reduced need
- **Long-Term Effectiveness:** By promoting natural balances, IPM methods can be effective over the long term, minimizing the risk of pest resistance.
- **Biodiversity Conservation:** IPM preserves beneficial organisms, supporting overall biodiversity within the orchard ecosystem.
- **Food Safety and Quality:** Reduced pesticide residues enhance the safety and quality of fruits for consumers.

Challenges and Considerations:

- **Knowledge and Training:** Successful IPM implementation requires knowledge of pest biology, monitoring techniques, and management strategies.
- **Adaptation:** IPM strategies need to be tailored to specific orchard conditions, pest species, and regional factors.
- **Continuous Monitoring:** Orchards need regular monitoring to promptly address pest outbreaks and implement appropriate control measures.

Integrated Pest Management embodies a harmonious approach to orchard management, fostering equilibrium between pest populations and natural controls. By prioritizing long-term sustainability, IPM offers a way forward that ensures healthy fruit production while respecting the delicate balance of the orchard ecosystem. Through informed decisions and the integration of diverse strategies, IPM demonstrates that productive orchards can flourish without compromising the environment or the quality of the harvest.

9

Disease Management in Fruit Crops: Preserving Orchard Health and Yield

Disease management is a critical aspect of fruit crop production, as diseases can significantly impact yield, fruit quality, and the overall health of orchards. Employing effective disease management strategies is essential to ensure a sustainable and productive fruit-growing operation. By integrating preventive measures, cultural practices, and targeted treatments, orchardists can mitigate disease outbreaks, minimize economic losses, and promote the long-term health of their fruit crops.

Preventive Measures:

- ✯ **Site Selection:** Choosing well-drained locations with proper air circulation reduces the likelihood of diseases caused by excessive moisture.
- ✯ **Resistant Varieties:** Planting disease-resistant or tolerant varieties can limit the impact of specific pathogens.
- ✯ **Sanitation:** Regularly removing diseased plant material, fallen leaves, and debris reduces the source of inoculum for pathogens.

Cultural Practices:

- ✯ **Proper Pruning:** Pruning to enhance sunlight penetration and airflow minimizes conditions favorable for disease development.
- ✯ **Adequate Spacing:** Providing sufficient spacing between plants prevents overcrowding, reducing humidity levels and disease spread.

- ✯ **Irrigation Management:** Avoiding over watering and utilizing drip irrigation instead of overhead methods can help prevent waterborne diseases.

Biological Control:

- ✯ **Beneficial Microorganisms:** Applying beneficial microbes that antagonize pathogens can suppress disease development.
- ✯ **Natural Predators:** Encouraging natural predators like ladybugs and predatory mites can help control pest insects that transmit diseases.

Chemical Control:

Fungicides play a crucial role in disease management by effectively combating fungal infections that can devastate fruit crops. These chemical compounds are designed to specifically target and eliminate pathogenic fungi, protecting the health and yield of orchards. While their use should be judicious and well-integrated within an overall disease management strategy, fungicides are valuable tools when other methods fall short.

Examples of Fungicides:

- ✯ **Triazoles:** Fungicides like tebuconazole and propiconazole inhibit fungal growth by disrupting the synthesis of essential components, ultimately preventing cell division and reproduction.
- ✯ **Strobilurins:** Compounds such as azoxystrobin and pyraclostrobin target the fungal mitochondria, leading to energy depletion and cell death.
- ✯ **Copper Compounds:** Copper-based fungicides, such as copper sulfate, act as protectants by creating a barrier on plant surfaces, preventing spore germination and infection.
- ✯ **Dithiocarbamates:** Fungicides like mancozeb and zineb work by inhibiting enzymes required for fungal growth, effectively controlling a range of diseases.
- ✯ **Benzimidazoles:** Thiophanate-methyl and carbendazim disrupt fungal cell division and growth by interfering with microtubule formation.
- ✯ **Quinone Outside Inhibitors (QoIs):** Fungicides like azoxystrobin and kresoxim-methyl target mitochondrial respiration, leading to fungal death.
- ✯ **DMI (*Demethylation Inhibitors*) Fungicides:** Tebuconazole and myclobutanil interfere with the production of sterols, essential components of fungal cell membranes.

Benefits of Fungicides:

- **Effective Disease Control:** Fungicides provide rapid and targeted control of fungal diseases, reducing crop losses.
- **Quick Response:** Fungicides can quickly address disease outbreaks and prevent further spread.
- **Precision:** Modern fungicides are often designed to target specific pathogens, minimizing harm to non-target organisms.
- **Enhanced Yield and Quality:** By preventing or mitigating disease damage, fungicides contribute to higher yields and better fruit quality.
- **Flexibility:** Fungicides can be tailored to the specific disease and its lifecycle, optimizing control strategies.

Integrated Disease Management (IDM): Holistic Disease Mitigation

Integrated Disease Management (IDM) is a comprehensive approach that seeks to harmonize various disease management strategies, including cultural practices, biological control, and judicious chemical use. IDM aims to optimize disease control while minimizing the impact on the environment and non-target organisms.

Components of IDM:

- **Disease Monitoring:** Regular monitoring helps identify disease trends, allowing timely interventions.
- **Cultural Practices:** Implementing practices like pruning, sanitation, and proper irrigation creates an environment less conducive to disease.
- **Biological Control:** Encouraging beneficial organisms can naturally suppress pathogen populations.
- **Chemical Control:** Fungicides are used strategically and selectively, minimizing their impact on non-target species.
- **Resistant Varieties:** Planting disease-resistant varieties reduces the susceptibility of crops to certain diseases.
- **Crop Rotation:** Rotating crops can disrupt disease cycles, preventing buildup of soil-borne pathogens.

Benefits of IDM:

- **Sustainability:** IDM reduces reliance on chemical treatments, promoting a balanced ecosystem.
- **Long-Term Effectiveness:** By combining various strategies, IDM reduces the risk of pathogen resistance.
- **Environmental Stewardship:** Minimizing chemical use and preserving beneficial organisms contribute to ecological health.

Challenges and Considerations:

- **Knowledge and Expertise:** Successful IDM requires a deep understanding of orchard ecosystems and the life cycles of pathogens.
- **Flexibility:** Orchards are dynamic systems, and IDM strategies may need adjustments based on changing conditions.

Disease management in fruit crops is a dynamic and multifaceted endeavor that requires a deep understanding of orchard ecosystems and the pathogens that threaten them. By adopting a holistic approach that integrates preventive measures, cultural practices, biological control, and targeted treatments, orchardists can safeguard the health of their fruit crops, ensuring sustainable production, high-quality fruits, and a thriving orchard ecosystem.

10

Organic Fruit Production: Nurturing Health & Sustainability from Orchard to Table

Organic fruit production represents a holistic approach to cultivating fruit crops that prioritize environmental health, sustainable practices, and the delivery of nutritious, flavorful fruits to consumers. Rooted in principles that avoid synthetic chemicals and foster natural processes, organic fruit production aligns with growing consumer demand for high-quality, environmentally friendly produce.

Organic Principles and Practices:

- **Soil Health:** Organic orchards prioritize soil fertility through composting, cover cropping, and reduced tillage, enhancing soil structure and nutrient availability.
- **Biological Diversity:** Encouraging diverse plant species, beneficial insects, and microorganisms promotes a balanced ecosystem that helps suppress pests and diseases.
- **Avoidance of Synthetic Chemicals:** Organic fruit production excludes synthetic pesticides, herbicides, and fertilizers, relying on natural alternatives instead.
- **Integrated Pest Management (IPM):** Organic orchards embrace IPM, incorporating cultural practices, biological controls, and organic-approved pesticides for disease and pest management.
- **Non-GMO Approach:** Organic fruit production prohibits the use of genetically modified organisms (GMOs), preserving genetic diversity.

Cultural Practices in Organic Orchards: Nurturing Health and Diversity

Cultural practices form the foundation of organic orchard management, fostering an environment that encourages the health of both fruit crops and the ecosystem as a whole. Rooted in ecological principles, these practices promote biodiversity, soil fertility, and natural pest management, all while minimizing the need for synthetic inputs. By adopting and adapting these techniques, organic orchardists create thriving ecosystems that produce flavorful and nutritious fruits while preserving the delicate balance of the natural world.

Crop Rotation and Diversity:

Crop rotation is a cornerstone of organic orchard cultural practices. Alternating fruit crops with other plants disrupts disease and pest cycles, preventing the buildup of specific pathogens and insects. This practice enhances soil fertility, as different crops utilize and replenish distinct sets of nutrients. Additionally, diverse plant species foster beneficial insects, creating a balanced ecosystem where pests are naturally controlled.

Cover Cropping and Mulching:

Cover cropping involves planting complementary plants between fruit trees to suppress weeds, improve soil structure, and enhance nutrient cycling. Leguminous cover crops fix nitrogen in the soil, benefiting neighboring fruit trees. Mulching with organic materials like straw or wood chips conserves soil moisture, moderates temperature, and provides habitat for beneficial organisms that contribute to natural pest control.

Companion Planting:

Companion planting involves strategically placing companion plants near fruit trees to harness their natural synergies. Some plants repel pests, while others attract pollinators or beneficial insects that prey on orchard pests. The interplay between diverse plant species creates a harmonious environment that reduces the need for chemical interventions.

Pruning and Canopy Management:

Pruning is a fundamental practice in organic orchards. Properly pruned trees allow sunlight to penetrate the canopy, reducing humidity and creating an environment less conducive to disease development. Well-ventilated canopies also aid in pest control, as they make the orchard less hospitable to certain insects.

Natural Pest Management:

Organic orchardists harness nature's own defenses to manage pests. By attracting beneficial insects like ladybugs and lacewings, orchards can naturally control pest populations. Additionally, planting trap crops diverts pests away from valuable fruit crops. These practices create a harmonious ecosystem where predator-prey interactions help maintain pest balance.

Irrigation and Water Management:

Efficient irrigation practices minimize water wastage and disease susceptibility. Drip irrigation directly targets root zones, reducing foliar moisture that can foster disease. Conserving water resources is not only environmentally responsible but also crucial for preventing waterborne diseases that thrive in damp conditions.

Cultural practices in organic orchards embody a philosophy of working with nature rather than against it. By enhancing biodiversity, promoting soil health, and fostering natural pest management, organic orchardists create thriving ecosystems that sustainably yield abundant, high-quality fruits. These practices reflect a holistic approach to agriculture, one that honors the interconnectedness of all living organisms and prioritizes the long-term health of the environment and its inhabitants.

Natural Pest Management:

Natural pest management is a cornerstone of sustainable agriculture, offering a holistic and environmentally friendly approach to controlling pests without relying on synthetic chemicals. Rooted in ecological principles, this practice leverages the natural balance of ecosystems to maintain pest populations at manageable levels. By harnessing the power of beneficial organisms and implementing a range of strategies, natural pest management minimizes crop damage, preserves biodiversity, and ensures the long-term health of agricultural systems.

Beneficial Organisms as Allies:

Beneficial insects, mites, birds, and other organisms serve as nature's allies in pest management. These predators and parasitoids prey on pest insects, acting as a front line defense in controlling their populations. Ladybugs, lacewings, and parasitic wasps are just a few examples of these natural predators that help maintain a harmonious balance in the agroecosystem.

Conservation and Augmentation:

Organic farmers actively encourage the presence of beneficial organisms through conservation and augmentation techniques. By providing shelter, nectar-rich plants, and habitat diversification, farmers create an environment that attracts and sustains these natural allies. In some cases, beneficial organisms can be introduced to bolster existing populations, particularly during critical pest infestations.

Trap Crops and Companion Planting:

Strategic planting of trap crops diverts pests away from valuable crops. These sacrificial crops attract pests, acting as a magnet and reducing damage to the main crops. Companion planting involves strategically intercropping pest-repellent plants with target crops, creating confusion for pests and inhibiting their colonization.

Pheromones and Biopesticides:

Pheromones are chemical signals emitted by insects to communicate with each other. Synthetic pheromones can be used to disrupt mating patterns and confuse pests, reducing their ability to reproduce. Biopesticides derived from naturally occurring organisms, such as bacteria or fungi, target specific pests while minimizing harm to non-target species.

Biological Control Agents:

Predatory insects, nematodes, and microorganisms can be introduced into the environment to target specific pests. For example, releasing parasitic nematodes into the soil can effectively control soil-dwelling pests, while introducing predator beetles can combat aphid infestations.

Cultural Practices that Promote Natural Pest Management

- **Crop Rotation:** Interrupts pest life cycles and reduces pest buildup.
- **Diverse Planting:** Increases habitat for beneficial insects and supports pollinators.
- **Habitat Enhancement:** Planting hedgerows, cover crops, and flowering plants creates shelter and food sources for natural predators.

Benefits of Natural Pest Management:

- ✭ **Environmental Sustainability:** Minimizes the use of synthetic chemicals, preserving biodiversity and ecosystem health.
- ✭ **Reduced Residues:** Decreases pesticide residues in food and minimizes risks to human health.
- ✭ **Long-Term Efficacy:** Promotes a balanced ecosystem, reducing the likelihood of pest resistance.
- ✭ **Cost-Effectiveness:** Lowers input costs by utilizing nature's pest control mechanisms.

Challenges and Considerations:

- ✭ **Complexity:** Implementing natural pest management requires a deep understanding of local ecosystems and pest dynamics.
- ✭ **Monitoring and Timing:** Regular monitoring is essential to detect pest outbreaks and deploy interventions at the right time.

Natural pest management is a testament to the intricate web of interactions that underpin our agricultural landscapes. By harnessing the power of beneficial organisms and embracing diverse strategies, farmers not only mitigate pest damage but also contribute to the health and resilience of the environment. This approach exemplifies the harmonious relationship between human activities and the natural world, offering a sustainable path forward for agriculture.

Organic Disease Management: Nurturing Orchard Health through Natural Balance

Organic disease management is a holistic approach to protecting fruit crops from diseases while adhering to the principles of sustainable and environmentally friendly agriculture. Rooted in the understanding that healthy plants have a natural resistance to diseases, organic disease management focuses on creating an ecosystem that supports plant vitality and discourages disease development. By integrating cultural practices, biological controls, and limited organic treatments, organic orchardists mitigate disease risks while fostering a balanced and resilient orchard ecosystem.

Resistant Varieties and Soil Health:

Selecting disease-resistant plant varieties forms the foundation of organic disease management. These varieties possess genetic traits that make them less susceptible to specific diseases. In addition, maintaining soil health through practices such as composting, cover cropping, and reduced tillage boosts the vigor of plants, enhancing their natural defenses against diseases.

Cultural Practices for Disease Prevention:

- ★ **Pruning and Canopy Management:** Proper pruning promotes good airflow and sunlight penetration, reducing humidity and creating conditions unfavorable for disease development.
- ★ **Sanitation:** Regular removal of diseased plant material and fallen leaves prevents the buildup of pathogens in the orchard.
- ★ **Crop Rotation:** Alternating fruit crops with other plants disrupts disease cycles and limits the buildup of pathogens in the soil.

Biological Controls:

- ★ **Beneficial Microorganisms:** Application of beneficial microorganisms, such as mycorrhizal fungi and bacteria, can enhance plant health and bolster their disease resistance.
- ★ **Antagonistic Organisms:** Introducing natural enemies of pathogens, such as certain bacteria and fungi, can suppress disease development.

Organic Treatments:

- ★ **Copper-Based Compounds:** Copper fungicides, approved for organic use, create a protective barrier on plant surfaces, inhibiting spore germination and disease spread.
- ★ **Sulfur:** Sulfur-based products act as fungicides and also have a mild insecticidal effect on some pests.
- ★ **Biological Products:** Biopesticides derived from naturally occurring organisms, such as Bacillus subtilis or Trichoderma, target specific diseases without harming beneficial organisms.

Integrated Disease Management (IDM):

Organic disease management embraces an integrated approach that synergizes various strategies. IDM combines cultural practices, biological controls, and approved organic treatments to optimize disease prevention and control.

Benefits of Organic Disease Management:

- ★ **Sustainability:** Organic disease management practices minimize the impact on the environment, preserving soil health and biodiversity.

- ✯ **Reduced Chemical Exposure:** Consumers and farmworkers are exposed to fewer synthetic chemicals through organic practices.
- ✯ **Resilient Ecosystems:** Promoting natural balance through organic practices reduces the likelihood of disease outbreaks and the need for intensive treatments.

Fertilization and Nutrient Management: Cultivating Orchard Health through Balanced Nutrition

Fertilization and nutrient management are fundamental components of successful orchard management, ensuring that fruit crops receive the essential elements they need for optimal growth, productivity, and fruit quality. Organic orchards prioritize a holistic approach that integrates natural nutrient sources, composting, cover cropping, and careful monitoring to maintain soil fertility and support the long-term health of fruit-bearing trees. By balancing nutrient inputs with the needs of the orchard ecosystem, organic orchardists create a sustainable foundation for bountiful harvests and thriving landscapes.

Understanding Nutrient Needs:

Each fruit crop has specific nutrient requirements that must be met for healthy growth and fruit development. Macro-nutrients like nitrogen, phosphorus, and potassium are crucial, while micro-nutrients like zinc, iron, and manganese are equally essential in smaller quantities. Organic orchardists assess soil nutrient levels and tailor fertilization plans to address any deficiencies.

Organic Nutrient Sources:

- ✯ **Composting:** Composting organic matter, such as plant residues and manure, enriches the soil with nutrients and promotes microbial activity that enhances nutrient availability.
- ✯ **Cover Crops and Green Manure:** Leguminous cover crops fix atmospheric nitrogen into the soil, benefiting subsequent fruit crops.
- ✯ **Mulching:** Organic mulches like straw or wood chips gradually release nutrients into the soil as they decompose.

Balancing Nutrient Inputs:

Organic orchardists prioritize nutrient management that aligns with the principles of sustainable agriculture. Over-fertilization can lead to nutrient imbalances, environmental pollution, and decreased soil health. Careful monitoring, soil testing, and adjusting fertilization based on plant growth stages help maintain a delicate nutrient equilibrium.

Avoiding Synthetic Fertilizers:

Organic orchards eschew synthetic fertilizers in favor of natural and organic nutrient sources. The use of synthetic chemicals contradicts the holistic principles of organic farming and may disrupt soil microbial communities and ecosystem balance.

Compost and Soil Health:

Organic matter-rich soil retains nutrients, improves soil structure, and enhances water-holding capacity. Compost application builds soil organic matter, providing a slow-release source of nutrients that nourishes plants over time.

Cover Cropping and Nitrogen Fixation:

Incorporating leguminous cover crops, such as clover or vetch, increases nitrogen availability in the soil. These plants have a symbiotic relationship with nitrogen-fixing bacteria that capture atmospheric nitrogen and make it accessible to other plants.

Challenges and Considerations:

- **Nutrient Imbalance:** Over-fertilization or inadequate nutrient management can lead to nutrient imbalances, affecting plant health and yield.
- **Crop-Specific Needs:** Different fruit crops have varying nutrient requirements that must be carefully tailored to their growth stages.

Fertilization and nutrient management in organic orchards epitomize the interconnectedness of soil health, plant vitality, and sustainable agriculture. By embracing natural nutrient sources, promoting composting and cover cropping, and exercising careful monitoring, organic orchardists cultivate a rich and balanced soil ecosystem that nourishes fruit-bearing trees. The result is not only healthy and flavorful fruits but also an enduring commitment to responsible land stewardship, where nutrient management aligns harmoniously with the cycles of nature.

Certification and Regulation: Ensuring Integrity in Organic Orchard Management

Certification and regulation are essential pillars of the organic industry, providing transparency, accountability, and consumer confidence in the production of organic fruits. Organic orchardists adhere to rigorous standards set forth by governing bodies to ensure that their practices align with the principles of sustainable agriculture and environmental stewardship. Through meticulous documentation, inspections, and adherence to prescribed guidelines, organic orchards uphold the integrity of organic farming and maintain the trust of consumers seeking healthful, responsibly produced fruits.

Organic Certification Process:

- **Application:** Orchardists interested in organic certification submit an application detailing their farming practices, inputs, and history of the land.
- **Documentation:** Detailed records of farming practices, inputs, soil amendments, and pest management strategies are maintained to demonstrate compliance with organic standards.
- **Inspection:** Certified organic farms undergo regular on-site inspections by accredited certification agencies. Inspectors verify that practices align with organic regulations and assess the overall integrity of the operation.
- **Transition Period:** Farms transitioning from conventional to organic practices must undergo a designated transition period to restore soil health and implement organic methods.

Regulatory Bodies:

- **National Organic Program (NOP):** In the United States, the NOP, part of the United States Department of Agriculture (USDA), establishes and enforces organic standards for agricultural products.
- **International Standards:** Many countries have their own organic certification programs that adhere to international standards established by organizations like the International Federation of Organic Agriculture Movements (IFOAM).

Organic Standards:

- **Prohibited Substances:** Organic regulations prohibit the use of synthetic pesticides, herbicides, and synthetic fertilizers. Approved substances must be naturally derived and listed in the National List of Allowed and Prohibited Substances.

- **Genetic Engineering and GMOs:** Organic standards prohibit the use of genetically modified organisms (GMOs) in all aspects of organic production.
- **Soil Health:** Organic regulations emphasize soil health and fertility through practices like composting, cover cropping, and reduced tillage.
- **Animal Welfare:** For organic livestock production, standards include providing access to pasture, ensuring animal well-being, and limiting the use of antibiotics and hormones.

Benefits of Certification and Regulation:

- **Consumer Confidence:** Organic certification assures consumers that the product meets stringent organic standards.
- **Market Access:** Certified organic products can access premium markets and command higher prices due to consumer demand for organically produced goods.
- **Environmental Protection:** Organic regulations promote practices that minimize environmental impact, conserve biodiversity, and support ecological balance.

Certification and regulation are the cornerstones that uphold the integrity of organic orchard management. By complying with rigorous standards and regulations, organic orchardists demonstrate their commitment to sustainable, responsible farming practices. These efforts not only produce healthful, environmentally friendly fruits but also contribute to a global movement that prioritizes the health of ecosystems, consumers, and the future of agriculture.

Benefits of Organic Fruit Production: Nurturing Health, Environment, and Sustainability

Organic fruit production offers a wealth of advantages that extend beyond the boundaries of the orchard. Rooted in sustainable practices and ecological harmony, organic fruit production not only yields flavorful and nutritious fruits but also contributes to the well-being of consumers, the environment, and future generations.

Environmental Preservation:

- ★ **Soil Health:** Organic practices prioritize soil health through composting, cover cropping, and reduced tillage, maintaining soil structure and preventing erosion.
- ★ **Biodiversity:** Organic orchards encourage diverse plant species, beneficial insects, and microorganisms, fostering a balanced ecosystem that naturally suppresses pests and diseases.
- ★ **Water Conservation:** Organic methods, such as mulching and efficient irrigation, minimize water wastage and prevent water pollution.
- ★ **Reduced Chemical Use:** Organic fruit production avoids synthetic pesticides and fertilizers, preventing the release of harmful chemicals into the environment.

Nutritional Quality:

- ★ **Enhanced Nutrients:** Organic fruits often have higher levels of certain nutrients and antioxidants due to enriched soil and reduced exposure to synthetic chemicals.
- ★ **No Residue Concerns:** Organic fruits are less likely to have pesticide residues, reducing potential health risks associated with conventional produce.

Flavor and AromA.

- ★ **Intense Flavors:** Consumers often report that organic fruits possess more intense flavors and aromas, attributed to healthier plants and nutrient-rich soils.

Human Health:

- ★ **Reduced Chemical Exposure:** Organic fruit production limits exposure to synthetic pesticides and herbicides for both consumers and farmworkers.
- ★ **Safe for Future Generations:** By avoiding synthetic chemicals, organic fruit production contributes to a healthier environment for future generations.

Sustainability:

- ★ **Long-Term Viability:** Organic practices promote soil health, biodiversity, and ecosystem balance, ensuring sustainable agriculture for years to come.

* **Resilient Systems:** Organic orchards are less prone to pest and disease outbreaks due to their emphasis on natural balance and integrated pest management.

Economic Impact:

* **Premium Markets:** Organic fruits often command higher prices in the market due to consumer preference for healthful and environmentally friendly products.
* **Local Economies**: Organic orchards often support local economies by creating jobs and promoting community engagement.

Social Responsibility:

* **Ethical Farming:** Organic fruit production emphasizes fair labor practices, animal welfare, and social responsibility.

Education and Awareness:

* **Consumer Awareness:** Organic fruit production raises consumer awareness about sustainable farming practices and fosters a deeper connection to the food they eat.
* **Inspiration for Change:** Successful organic orchards can inspire other growers to adopt more sustainable practices, contributing to broader positive changes in agriculture.

The benefits of organic fruit production extend far beyond the orchard gates. By prioritizing environmental health, sustainable practices, and consumer well-being, organic orchardists contribute to a global movement that seeks to harmonize agriculture with nature. Organic fruits not only satisfy taste buds but also serve as a symbol of responsible land stewardship, nurturing landscapes for present and future generations while providing a tangible link between consumers and the vital processes of food production.

Challenges and Considerations in Organic Fruit Production: Navigating the Path of Sustainability

While organic fruit production offers numerous benefits, it also presents certain challenges and considerations that require thoughtful planning and proactive management. These challenges range from agronomic complexities to market dynamics, necessitating a balanced approach that takes into account the intricacies of organic farming and the broader agricultural landscape.

Transition Period:

- **Soil Restoration:** Converting conventional orchards to organic practices requires time for soil health and fertility to be restored, impacting immediate yields.
- **Ecosystem Balance:** Establishing natural predator-prey relationships and overall ecosystem balance may take time to fully develop.

Pest and Disease Management:

- **Labor Intensity:** Organic pest and disease management can be more labor-intensive, requiring vigilant monitoring and implementation of natural control measures.
- **Risk of Outbreaks:** Without conventional chemical interventions, there may be periods of increased pest and disease pressure that need to be carefully managed.

Market Demand and Pricing:

- **Fluctuating Demand:** Organic fruit demand can vary seasonally and geographically, impacting market access and prices.
- **Premium Prices:** While organic fruits often command higher prices, orchardists must weigh potential returns against increased production costs.

Transitioning to Organic Practices:

- **Knowledge and Training:** Shifting from conventional to organic practices requires a deep understanding of organic principles and adaptation of management techniques.
- **Financial Investment:** Initial costs associated with organic certification, infrastructure upgrades, and soil enhancement can be substantial.

Weed Management:

- **Labor and Timing:** Organic weed management, often relying on manual labor and mechanical methods, requires careful planning and execution.
- **Competing with Weeds:** Weeds can compete with fruit crops for water and nutrients, impacting yields if not effectively managed.

Diverse Ecosystem Challenges:

- **Complexity:** Managing a diverse ecosystem, including beneficial and pest species, requires a nuanced understanding of their interactions.

- ✯ **Learning Curve:** Orcharding practices need to be adapted to each unique ecosystem, and there may be a learning curve for orchardists.

Market Access and Consumer Education:

- ✯ **Certification Process:** Meeting organic certification requirements can be demanding, involving meticulous record-keeping and compliance.
- ✯ **Consumer Awareness:** Educating consumers about the value and principles of organic fruit production can influence market demand.

Climate and Weather Variability:

- ✯ **Adapting to Climate Change:** Organic orchards may face challenges from changing weather patterns, requiring adaptation strategies.

The challenges and considerations in organic fruit production reflect the dynamic nature of sustainable agriculture. While embracing organic practices brings numerous benefits, orchardists must navigate complex interactions within ecosystems, market dynamics, and the inherent uncertainties of farming.

Vegetables

1

Introduction to Vegetable Science

The Significance of Vegetables in Human Nutrition

Vegetables are the vibrant, nutrient-rich jewels of the plant world that play a crucial role in human nutrition. Packed with vitamins, minerals, fiber, and phytochemicals, vegetables offer an array of health benefits that contribute to overall well-being. They are a prime source of essential nutrients like vitamin C, vitamin A, potassium, and folate, which support various bodily functions, including immune health, vision, and heart function. Additionally, the dietary fiber found in vegetables aids digestion, regulates blood sugar levels, and promotes a healthy gut microbiome.

The nutritional composition of different vegetables, highlighting their individual strengths and the synergistic effects of consuming a diverse range of plant-based foods. It will also explore the concept of "eating the rainbow" – the idea that consuming a variety of colorful vegetables ensures a balanced intake of nutrients and antioxidants.

The Cultural Tapestry of Vegetables in Culinary Traditions

Beyond their nutritional value, vegetables are woven into the fabric of culinary traditions around the world. From the fiery curries of India to the delicate art of French cuisine, vegetables serve as versatile and essential ingredients that define the flavors and textures of various dishes. They provide a canvas for culinary creativity, enabling chefs and home cooks alike to experiment with flavors, techniques, and presentations.

The cultural significance of vegetables in different cuisines, showcasing how various vegetables are celebrated, prepared, and enjoyed across the globe. It will touch upon the traditional practices of fermenting, pickling, and preserving vegetables to extend their availability and enhance their flavors.

The Journey from Farm to Table: Understanding Vegetable Supply Chains

The journey of vegetables from the farm to our tables involves a complex web of processes that encompass cultivation, harvesting, distribution, and consumption. This intricate supply chain highlights the interdependence of various stakeholders, including farmers, distributors, retailers, and consumers. As we explore this journey, we uncover the challenges and innovations that shape the availability, quality, and affordability of vegetables.

The Historical Evolution of Vegetable Cultivation and Consumption

The history of vegetable cultivation and consumption is a captivating narrative that traces humanity's journey from foraging to sophisticated agricultural practices. Through centuries of trial and error, experimentation, and cross-cultural exchange, humans have transformed wild plants into the diverse array of cultivated vegetables we enjoy today.

Origins of Vegetable Cultivation

The origins of vegetable cultivation can be traced back to the dawn of agriculture, when humans transitioned from a nomadic, hunter-gatherer lifestyle to settled farming communities. The fertile crescent of Mesopotamia, often referred to as the "cradle of agriculture," witnessed the cultivation of early vegetables such as lentils, chickpeas, and peas around 10,000 years ago. These pioneering efforts marked the beginning of humanity's deliberate manipulation of plant species for sustenance.

Ancient Civilizations and Horticultural Advances

As civilizations flourished across different regions, vegetable cultivation became an integral part of agricultural systems. In ancient Egypt, records indicate the cultivation of a variety of vegetables, including lettuce, onions, and cucumbers, along the banks of the Nile River. Similarly, the Greeks and Romans embraced horticulture, with writings from scholars like The ophrastus detailing cultivation practices for vegetables like carrots, radishes, and cabbages.

The Medieval Period and Renaissance

The Middle Ages witnessed both continuity and transformation in vegetable cultivation. Monastic gardens preserved and cultivated various vegetables, contributing to their availability throughout Europe. The Renaissance period saw renewed interest in botany, leading to the discovery of new vegetable species through exploration and trade. This era marked the exchange of crops between the Old World and the New World, known as the Columbian Exchange, introducing tomatoes, potatoes, and peppers to European diets.

Industrialization and Modern Agriculture

The advent of industrialization and technological advancements revolutionized agriculture, including vegetable cultivation. Improved transportation, irrigation systems, and seed breeding techniques led to increased productivity and greater variety in vegetables. The development of modern agricultural science further accelerated progress, enabling the creation of hybrid vegetable varieties with enhanced qualities such as disease resistance and higher yields.

Contemporary Trends in Vegetable Consumption

In recent decades, the importance of vegetables in human health and nutrition has gained widespread recognition. The rise of vegetarianism, veganism, and plant-based diets has led to a reevaluation of the role of vegetables in modern diets. With a growing emphasis on sustainable food systems, local and organic vegetable cultivation has gained prominence, fostering a deeper connection between consumers and the sources of their food.

Importance and Classification of Vegetables

Vegetables, often referred to as the cornerstone of a healthy diet, occupy a pivotal place in human nutrition, culinary culture, and agriculture. This chapter delves into the multifaceted significance of vegetables and their diverse classification based on botanical characteristics, shedding light on the incredible diversity that these plants bring to our lives.

At the heart of human nutrition lies the profound importance of vegetables. These edible plant parts provide an abundance of vitamins, minerals, antioxidants, and dietary fiber that are essential for maintaining optimal health. From bolstering the immune system to supporting proper digestion and reducing the risk of chronic diseases such as heart disease and diabetes, the nutritional profile of vegetables is a treasure trove of wellness.

Beyond their nutritional contributions, vegetables are culinary stars that elevate the sensory experience of food. Their vibrant colors, textures, and flavors not only entice the palate but also serve as essential components of countless dishes. Whether it's the crispness of a fresh salad, the hearty warmth of a vegetable stew, or the delicate balance they lend to gourmet creations, vegetables are culinary chameleons that can be transformed into an array of gastronomic delights.

In the realm of agriculture, the cultivation of vegetables plays a crucial role in sustainable food systems. Unlike resource-intensive animal agriculture, vegetable cultivation requires less land, water, and energy. Furthermore, the practice of crop rotation involving vegetables can naturally enrich soil fertility and deter pests. The efficient use of resources in vegetable cultivation aligns with the principles of environmental stewardship and contributes to a more resilient and ecologically balanced food production.

Classification of Vegetables

The diversity among vegetables is astounding, encompassing a wide range of plant species with distinct characteristics and uses. Classification based on botanical attributes provides insight into their relationships and helps us understand their growth patterns, requirements, and traits.

One common classification divides vegetables into different categories based on the edible part of the plant. Root vegetables, such as carrots and potatoes, grow underground and store essential nutrients. Stem vegetables, including asparagus and celery, feature edible above-ground stems that provide unique textures and flavors. Leafy vegetables, exemplified by spinach and lettuce, offer an array of vitamins and minerals. Meanwhile, flower vegetables like broccoli and cauliflower showcase the edible blossoms of certain plants.

Fruits, in botanical terms, are the mature ovaries of plants that contain seeds. Surprisingly, many foods that we consider vegetables are, botanically speaking, fruits. Tomatoes, peppers, and cucumbers, for instance, are all fruits according to botanical classification. This discrepancy between culinary and botanical definitions underscores the complexity of vegetable classification and the cultural nuances that influence how we perceive them.

The botanical grouping of vegetables extends further into families, orders, and genera, reflecting their evolutionary relationships. Understanding these classifications is not only important for horticulturists and scientists but also informs agricultural practices and breeding efforts aimed at developing improved vegetable varieties.

Classification of Vegetables: Unveiling Nature's Diversity

The world of vegetables is a rich tapestry of diverse forms, flavors, and functions. In this chapter, we embark on a journey to uncover the intricate classification systems that help us make sense of this remarkable diversity. By understanding the botanical relationships that define different types of vegetables, we gain insights into their growth, cultivation, and unique attributes.

Taxonomic Classifications

Taxonomy, the science of classification, offers a systematic approach to categorizing living organisms based on their evolutionary relationships. Within the plant kingdom, vegetables are no exception. Botanists utilize taxonomy to arrange vegetables into hierarchical groups that reflect shared characteristics and ancestry.

At the broadest level, vegetables belong to the kingdom Plantae. This diverse group encompasses an array of life forms, from towering trees to delicate herbs. Moving down the hierarchy, we encounter divisions into phyla, classes, and orders, each step revealing more about the plant's evolutionary history and features.

The Edible Parts: A Classification Approach

One of the simplest classification approaches is based on the edible parts of vegetables. This categorization encompasses several distinct groups that highlight the variety of ways vegetables contribute to our diets and culinary creations.

Root vegetables, such as carrots, radishes, and sweet potatoes, derive their nutrition from below the soil's surface. These underground treasures store vital nutrients, offering sustenance in diverse cultures around the world.

Leafy vegetables, like spinach, lettuce, and kale, present a spectrum of textures and flavors. Their nutrient-dense leaves contribute vitamins, minerals, and antioxidants to our diets.

Stem vegetables, including celery and asparagus, offer a unique twist to culinary creations. These crisp and flavorful parts of plants showcase the versatility of vegetables in our meals.

Flower vegetables, exemplified by broccoli and cauliflower, challenge our perception of edible parts. These flower buds deliver a burst of flavor and nutrients, providing a creative twist to our dining experiences.

Fruits, in botanical terms, are mature ovaries containing seeds. Despite culinary distinctions, many vegetables, like tomatoes, peppers, and eggplants, are botanically classified as fruits.

Beyond the Basics: Genetic and Hybrid Classifications

In modern times, advances in genetics have led to another dimension of vegetable classification. Genetic markers and DNA analysis enable scientists to delve into the genetic relationships between different vegetable varieties and species. Hybrid vegetables, the result of deliberate crossbreeding, further expand the array of choices available to consumers. These hybrids often combine the best traits of different vegetable varieties, enhancing qualities such as disease resistance, yield, and flavor.

Mulching and Cropping Systems in Vegetable Cultivation

In the realm of vegetable cultivation, successful farming practices are essential for maximizing yields, conserving resources, and ensuring sustainable agricultural practices.

Mulching: A Shield for Soil and Plants

Mulching, the practice of covering the soil surface with a protective layer, is a technique with far-reaching benefits in vegetable cultivation. Mulches come in various forms, including organic materials like straw, leaves, and compost, as well as synthetic materials like plastic films. Their primary function is to create a protective barrier between the soil and the external environment. One of the key advantages of mulching is its ability to conserve soil moisture. By reducing evaporation and minimizing water loss, mulches help ensure a consistent supply of water to vegetable plants. This is particularly crucial in arid and semi-arid regions where water scarcity poses a challenge to agriculture. Mulches also play a role in weed suppression, a vital aspect of sustainable farming. By blocking sunlight and inhibiting weed growth, mulches minimize competition for water, nutrients, and light, contributing to healthier vegetable crops.

Cropping Systems: Strategic Arrangements for Success

The arrangement of crops within a field, known as cropping systems, influences factors such as yield, pest management, and soil health. Different cropping systems are designed to optimize land use, minimize risk, and improve overall efficiency in vegetable cultivation. One common cropping system is monoculture, where a single vegetable crop is grown in a field. While this approach may simplify management, it can also increase the vulnerability of crops to diseases and pests. Crop rotation, a traditional practice, involves planting different crops in succession to break pest and disease cycles and enhance soil fertility. Intercropping, another cropping strategy, involves planting two or more crops in close proximity. This practice promotes biodiversity, reduces the spread of pests and diseases, and optimizes the use of resources such as sunlight and water. Agroforestry, a more complex cropping system, combines trees or shrubs with agricultural crops. This approach enhances soil fertility, provides shade, and offers additional products like fruits and timber.

Synergy Between Mulching and Cropping Systems

The synergy between mulching and cropping systems is a potent strategy for achieving sustainable vegetable cultivation. When used together, these practices enhance each other's benefits, amplifying the positive impacts on soil health, water conservation, and crop productivity.

2

Vegetable Planting and Spacing

Proper vegetable planting and spacing are crucial for healthy plant growth, optimal yields, and disease prevention. Each vegetable has specific requirements for how deep to plant the seeds or seedlings and how much space they need to grow. Here's a general guideline for vegetable planting and spacing:

1. Site Preparation:

- ✯ Prepare the soil by removing weeds, rocks, and debris. Amend the soil with compost or organic matter to improve fertility and drainage.

2. Seed Planting:

- ✯ Follow the planting depth guidelines on the seed packet. As a general rule, plant seeds at a depth of about two to three times their diameter.
- ✯ Larger seeds are planted deeper than smaller seeds.
- ✯ Thin the seedlings if they emerge too closely, ensuring proper spacing for healthy growth.

3. Seedling Transplanting:

- ✯ Transplant seedlings when they have developed a few true leaves and are strong enough to handle transplanting.
- ✯ Dig a hole slightly larger than the root ball of the seedling and gently place it in the hole. Fill in with soil and water well.

4. Spacing Guidelines:

- ✯ **Leafy Greens (Lettuce, Spinach, Kale):** Plant 4 to 12 inches apart, depending on the mature size of the plant.

- ★ **Root Vegetables (Carrots, Radishes, Beets):** Plant 2 to 4 inches apart. Thin as needed to achieve proper spacing.
- ★ **Bush Beans:** Plant 3 to 6 inches apart in rows spaced 18 to 24 inches apart.
- ★ **Pole Beans:** Plant 4 to 6 inches apart at the base of the trellis or support.
- ★ **Tomatoes:** Transplant at least 18 to 36 inches apart in rows spaced 24 to 36 inches apart.
- ★ **Peppers:** Transplant 18 to 24 inches apart in rows spaced 24 to 36 inches apart.
- ★ **Cucumbers:** Plant 12 to 24 inches apart in rows spaced 36 to 60 inches apart.
- ★ **Squash (Zucchini, Summer Squash):** Plant 18 to 36 inches apart in rows spaced 36 to 60 inches apart.
- ★ **Broccoli and Cauliflower:** Plant 18 to 24 inches apart in rows spaced 24 to 36 inches apart.
- ★ **Corn:** Plant seeds in blocks rather than rows, with plants spaced 12 to 18 inches apart.

5. Companion Planting:

- ★ Consider companion planting to maximize space and enhance plant health. Some plants thrive when grown together, while others may inhibit each other's growth.

6. Succession Planting:

- ★ Plan for multiple plantings throughout the growing season to extend the harvest period.

7. Vertical Gardening:

- ★ Utilize vertical space by planting vining crops on trellises, allowing you to grow more in limited space.

8. Containers:

- ★ Adjust spacing based on the container size. Follow the recommendations on the seed packet or plant label for container gardening.

Remember that specific spacing requirements can vary depending on factors such as plant variety, climate, and local growing conditions. It's important to research the recommended spacing for each vegetable you plan to grow to ensure optimal results and healthy plants.

Vegetable Seed Production

Tomato (*Solanum lycopersicum*):

- **Selection of Parent Plants:** Choose parent plants with desired traits such as fruit size, color, disease resistance, and flavor.
- **Isolation and Pollination:** Isolate varieties to prevent cross-pollination. Pollination can be achieved through self-pollination or controlled hand-pollination.
- **Fermentation:** To remove the gel around tomato seeds, ferment the seeds in water for a few days, stirring occasionally. Rinse and dry the seeds after fermentation.
- **Drying and Extraction:** Spread the seeds on a tray to dry, then separate them from the dried fruit material. Remove any remaining debris.
- **Cleaning and Sorting:** Clean and sort the seeds to remove any remaining impurities or damaged seeds.
- **Germination Testing:** Conduct germination tests to assess seed viability.
- **Packaging and Storage:** Package the seeds in airtight containers, label them with variety information, and store in a cool, dry place.

Pepper (*Capsicum* spp.):

- **Parent Plant Selection:** Choose parent plants based on fruit size, shape, color, and flavor.
- **Isolation and Pollination:** Isolate varieties to avoid cross-pollination between different types of peppers. Hand-pollinate flowers to ensure controlled pollination.
- **Fermentation (Optional):** Fermentation is not usually necessary for pepper seeds.
- **Drying and Extraction:** Dry the seeds thoroughly on a paper towel, then extract them from the dried fruits.
- **Cleaning and Sorting:** Clean and sort the seeds to remove any chaff or debris.
- **Germination Testing:** Test seed germination to assess viability.
- **Packaging and Storage:** Package the seeds in labeled containers and store them in a cool, dry place.

Cucumber (*Cucumis sativus*):

- ★ **Parent Plant Selection:** Choose parent plants with desired fruit characteristics, disease resistance, and yield.
- ★ **Isolation and Pollination:** Isolate varieties to prevent cross-pollination. Hand-pollinate cucumber flowers.
- ★ **Drying and Extraction:** Dry the seeds on paper towels, then extract them from the dried fruits.
- ★ **Cleaning and Sorting:** Clean and sort the seeds to remove any debris.
- ★ **Germination Testing:** Test seed germination for viability.
- ★ **Packaging and Storage:** Package the seeds in labeled containers and store in a cool, dry place.

Carrot (*Daucus carota*):

- ★ **Parent Plant Selection:** Choose parent plants with desired root shape, size, and color.
- ★ **Isolation and Pollination:** Isolate carrot varieties to prevent cross-pollination. Carrots are typically biennial plants that flower in the second year.
- ★ **Drying and Extraction:** Allow the seeds to mature on the plant, then dry the umbels (seed heads) in a well-ventilated area. Remove the seeds from the umbels.
- ★ **Cleaning and Sorting:** Clean and sort the seeds to remove debris and empty seeds.
- ★ **Germination Testing:** Test seed germination to assess viability.
- ★ **Packaging and Storage:** Package the seeds in labeled containers and store them in a cool, dry place.

Lettuce (Lactuca sativa):

- ★ **Parent Plant Selection:** Choose parent plants with desired leaf characteristics and disease resistance.
- ★ **Isolation and Pollination:** Isolate lettuce varieties to prevent cross-pollination. Lettuce is primarily self-pollinating.
- ★ **Drying and Extraction:** Allow the seeds to mature on the plant, then cut the seed heads and place them in a paper bag to dry further. Shake out the seeds.

- **Cleaning and Sorting:** Clean and sort the seeds to remove debris.
- **Germination Testing:** Test seed germination to assess viability.
- **Packaging and Storage:** Package the seeds in labeled containers and store in a cool, dry place.

3

Organic Vegetable Gardening

Organic vegetable gardening involves cultivating vegetables using natural and sustainable methods without the use of synthetic fertilizers, pesticides, or genetically modified organisms. It focuses on nurturing the health of the soil, plants, and environment while producing nutritious and safe food. Here's a detailed guide to organic vegetable gardening:

1. Site Selection:

- ✯ Choose a location with ample sunlight (at least 6-8 hours per day), good air circulation, and well-drained soil.
- ✯ Consider factors like proximity to water sources and protection from strong winds.

2. Soil Preparation:

- ✯ Test the soil to determine its pH and nutrient content.
- ✯ Amend the soil with compost, well-rotted manure, and other organic matter to improve soil structure, fertility, and water retention.

3. Crop Selection:

- ✯ Choose vegetable varieties suitable for your climate, soil type, and growing season.
- ✯ Consider crop rotation to prevent soil-borne diseases and nutrient depletion.

4. Companion Planting:

- ✯ Plant companion crops that benefit each other, such as planting marigolds to deter pests or using tall plants to provide shade for smaller ones.

5. Seed Starting and Transplanting:

- ✯ Start seeds indoors or purchase organic seedlings.
- ✯ Harden off seedlings by gradually exposing them to outdoor conditions before transplanting.

6. Planting:

- ✯ Plant seeds or seedlings at the appropriate depth and spacing, following planting guides on seed packets.
- ✯ Water newly planted vegetables thoroughly.

7. Organic Fertilization:

- ✯ Use organic fertilizers such as compost, well-rotted manure, bone meal, and fish emulsion.
- ✯ Apply fertilizers based on soil test recommendations to ensure balanced nutrient levels.

8. Mulching:

- ✯ Apply organic mulch (straw, leaves, compost) around plants to retain soil moisture, suppress weeds, and improve soil structure.

9. Pest Management:

- ✯ Encourage beneficial insects by providing habitat and nectar sources.
- ✯ Use physical barriers like row covers to prevent pest access.
- ✯ Employ companion planting and crop rotation to reduce pest populations.
- ✯ Use natural remedies like neem oil, garlic spray, and insecticidal soap to control pests.

10. Disease Prevention:

- ✯ Choose disease-resistant varieties whenever possible.
- ✯ Practice good sanitation by removing diseased plant material.
- ✯ Ensure proper spacing between plants for air circulation.

11. Watering:

- ✯ Water deeply and consistently, preferably in the morning, to reduce disease risk and maximize water absorption.

12. Weeding:

- ✯ Regularly remove weeds to prevent competition for nutrients and water.

13. Harvesting:

- ✯ Harvest vegetables at their peak ripeness for the best flavor and nutritional content.
- ✯ Use proper harvesting techniques to avoid damaging plants.

14. Soil Health Maintenance:

- ✯ Continuously add organic matter to the soil through composting and cover cropping.
- ✯ Avoid compacting the soil by minimizing foot traffic in the garden.

15. Record Keeping:

- ✯ Keep a garden journal to track planting dates, varieties, pest and disease issues, and overall garden performance.

Organic vegetable gardening promotes sustainability, biodiversity, and healthy ecosystems. By focusing on soil health, natural pest control, and responsible cultivation practices, you can produce bountiful and nutritious crops while minimizing the environmental impact of your garden.

4

Weed Management Techniques in Vegetable Crops

Weeds are unwanted plants that compete with vegetable crops for essential resources, such as nutrients, water, sunlight, and space. Effective weed management is crucial for maintaining optimal vegetable crop yields and quality. This chapter discusses various weed management techniques and their applications in vegetable crops, along with references to support the presented information.

Cultural Weed Management Techniques

Crop Rotation

Crop rotation is a well-established cultural practice that involves planting different crops in a specific sequence on the same piece of land. This technique helps disrupt weed life cycles by altering the environmental conditions that favor particular weed species. Crop rotation can reduce weed populations and minimize the buildup of weed seed banks (Légère et al., 2018).

Cover Crops

Cover crops are non-cash crops planted between main crop seasons to cover the soil and suppress weed growth. They compete with weeds for resources and create a physical barrier that inhibits weed germination and growth. Leguminous cover crops can also fix nitrogen, providing nutrients for the subsequent vegetable crop (Teasdale & Mohler, 2000).

Mechanical Weed Management Techniques

Mulching

Mulching involves covering the soil surface with organic or synthetic materials to suppress weed growth. Organic mulches, such as straw or wood chips, can physically block sunlight from reaching weed seedlings, preventing their establishment. Synthetic mulches, like black plastic, create a barrier that inhibits weed growth while also warming the soil and enhancing crop growth

Mechanical Cultivation

Mechanical cultivation includes activities like plowing, harrowing, and hoeing to disturb the soil and uproot weeds. Timely cultivation can disrupt weed growth and bury weed seedlings, reducing their competitive advantage. However, care should be taken to avoid damaging vegetable crops during cultivation .

Chemical Weed Management Techniques

Herbicides

Herbicides are chemical compounds designed to selectively control weeds. Pre-emergence herbicides are applied before weed emergence, while post-emergence herbicides are applied after weed seedlings have emerged. Careful selection and application of herbicides are essential to minimize damage to vegetable crops and non-target species .

Integrated Weed Management

Integrated Weed Management (IWM) involves combining multiple weed control strategies to maximize their effectiveness while minimizing negative impacts on the environment and crop production. A holistic approach that considers factors such as crop rotation, cultural practices, mechanical methods, and judicious herbicide use can lead to sustainable weed management in vegetable crops .

Effective weed management is crucial for successful vegetable crop production. Cultural, mechanical, and chemical weed management techniques, along with integrated approaches, offer a range of options to address weed challenges. A combination of these techniques tailored to specific crop and environmental conditions can lead to improved weed control and enhanced vegetable crop yields.

5

Cultivation of Major Vegetable Crops

Onions (*Allium cepa*) hold profound significance as a culinary staple and as a vital crop in agricultural cultivation. In the culinary realm, onions are indispensable, adding depth, flavor, and aroma to a wide array of dishes across diverse cuisines. Their versatility allows them to play starring and supporting roles in everything from soups and stews to salads and sauces. Onions, whether caramelized, sautéed, or eaten raw, serve as a fundamental ingredient that transforms ordinary recipes into flavorful culinary experiences.

Beyond their culinary allure, the agricultural importance of onions is equally noteworthy. As a cool-season crop, onions are cultivated in various regions, contributing to global food security. Their role in crop rotation strategies is significant, as onions help break pest and disease cycles for subsequent crops. Moreover, onions exhibit certain natural repellent properties that deter pests, further enhancing their role in integrated pest management systems.

The economic impact of onion cultivation is substantial. As a widely consumed and traded commodity, onions play a pivotal role in local economies and international trade. Their long shelf life after harvest and suitability for storage make them an economically viable option for both small-scale and large-scale farmers. Furthermore, the relative simplicity of onion cultivation and the relatively short growing cycle contribute to its appeal for farmers seeking reliable income. From a nutritional standpoint, onions offer health benefits as well. They are rich in antioxidants, vitamins, and minerals, contributing to overall well-being. The presence of compounds like quercetin in onions is associated with potential anti-inflammatory and immune-boosting properties.

Cultivating onions involves a series of steps and practices that ensure healthy growth, bulb development, and a successful harvest. Here's a general overview of the cultivation process for onion crops:

Site Selection:

- ★ Choose a well-drained field with good sunlight exposure. Onions require at least 6-8 hours of direct sunlight per day.
- ★ The soil should be fertile, loamy, and rich in organic matter. Soil pH should ideally be between 6.0 and 7.5.

Variety Selection:

- ★ Choose onion varieties based on local climate, market demand, and intended use (fresh consumption, storage, or processing).
- ★ Common onion varieties include 'Yellow Globe', 'Red Creole', 'Walla Walla Sweet', 'Texas Early Grano', and more.

Soil Preparation:

- ★ Begin soil preparation by plowing or tilling the field to break up clods and create a fine seedbed.
- ★ Incorporate well-rotted compost or organic matter to improve soil structure and fertility.

Sowing or Planting:

- ★ Onions can be grown from seeds or sets (small bulbs).
- ★ Sow onion seeds in rows with appropriate spacing, typically 4-6 inches apart.
- ★ Alternatively, plant onion sets at a depth of about 1 inch and spaced according to variety recommendations.

Fertilization:

- ★ Conduct soil testing to determine nutrient requirements.
- ★ Apply a balanced fertilizer rich in phosphorus and potassium, as onions require these nutrients for bulb development.
- ★ Nitrogen is crucial for leaf growth, but excessive nitrogen can delay bulb formation.

Mulching:

- ★ Apply mulch to conserve soil moisture, regulate temperature, and suppress weeds.

- Common mulch materials include straw, hay, or plastic sheets.

Irrigation:

- Onions require consistent moisture throughout their growth cycle.
- Use drip irrigation or soaker hoses to deliver water directly to the root zone and reduce water wastage.

Weed Control:

- Keep the onion beds weed-free to prevent competition for nutrients and space.
- Hand weeding, hoeing, mulching, or herbicides can be used for weed control.

Thinning and Transplanting (if needed):

- If you started with onion sets, thin out overcrowded plants to the recommended spacing.
- Transplant seedlings if started indoors into the prepared field with proper spacing.

Pest and Disease Management:

- Monitor for pests like onion thrips, aphids, and onion maggots. Use appropriate insecticides if needed.
- Practice crop rotation and remove infected plants to manage diseases like onion downy mildew and white rot.

Bulb Formation:

- As onions grow, they form bulbs in response to day length.
- Some varieties form bulbs earlier (short-day onions), while others require longer daylight (long-day onions).

Harvesting:

- Harvest onions when the tops start to yellow and fall over.
- Gently lift the bulbs from the soil, and allow them to cure in a dry, well-ventilated area for a couple of weeks.

Storage:

- After curing, remove the outer dry scales and trim the tops to about 1 inch.

- ★ Store onions in a cool, dry place with good ventilation to prevent mold and sprouting.

Garlic

Garlic (*Allium sativum*) holds multifaceted significance, encompassing both culinary allure and agricultural benefits. In the culinary world, garlic's aromatic and robust flavor serves as a cornerstone of countless dishes across diverse cuisines, elevating taste profiles with its distinctive notes. From aromatic stir-fries to hearty stews, garlic's culinary contributions are undeniable, weaving a tapestry of flavors that delight palates globally. Beyond its culinary charm, garlic has long been esteemed for its potential health benefits, including its reputed immune-boosting properties and positive effects on cardiovascular health.

However, garlic's importance extends well into the realm of agriculture. This humble bulb is remarkably adaptable, thriving in various climates and soil types, rendering it an accessible crop for a wide range of regions. Its cultivation plays a pivotal role in crop rotation strategies, effectively mitigating pests and diseases that afflict subsequent crops. Garlic's extracts also serve as potent biopesticides, aligning with sustainable and organic farming practices. Beyond protection, garlic extracts induce resistance in neighboring plants, fostering a more resilient and balanced agricultural ecosystem.

The health benefits attributed to garlic find relevance not only in human well-being but also in plant vitality. Its natural antimicrobial properties hold promise for warding off pathogens in both medical and agricultural contexts. As a crucial element in cultural traditions and rituals, garlic symbolizes its rich historical and cross-cultural importance.

Economically, garlic's impact reverberates through global trade and local markets. It is a sought-after commodity, contributing to economic growth and livelihoods in producing regions. The crop's profitability is complemented by its relatively low input requirements, offering an attractive option for farmers seeking viable ventures. Moreover, garlic's extended shelf life after proper curing and storage adds to its commercial appeal, ensuring a reliable supply even beyond harvest.

Cultivation Practices:

Soil Preparation:

- ★ Select a well-drained, fertile soil with a pH range of 6.0 to 7.5.

- Incorporate organic matter like compost or well-rotted manure to improve soil structure and nutrient content.

Planting Time:

- Garlic is typically planted in the fall, a few weeks before the first frost date in your area.
- This allows the garlic to establish roots before winter.

Varieties

There are several varieties of garlic, each with distinct flavors, bulb types, and growth characteristics. Here are descriptions of some popular garlic varieties:

Softneck Garlic (*Allium sativum* var. *sativum*):

- Soft neck garlic is one of the most common types and is known for its pliable stems and flexibility.
- It typically has multiple layers of cloves arranged in concentric circles around a central stem.
- Varieties within the softneck category include:
  - **Silverskin Garlic:** A versatile and strong-flavored variety with white bulbs and papery skin. It stores well and is commonly used for braiding.
  - **Artichoke Garlic:** This variety produces large, tightly packed bulbs with easy-to-peel cloves. It has a milder flavor and is suitable for roasting and sautéing.

Hardneck Garlic (Allium sativum var. *ophioscorodon*):

- Hardneck garlic is characterized by its stiff central stem, or "scape," which produces a flower head called a "scape."
- It typically has fewer, larger cloves arranged in a single layer around the stem.
- Varieties within the hardneck category include:
  - **Rocambole Garlic:** Known for its rich, complex flavor and easy-to-peel cloves. The cloves are often arranged in a spiral pattern. It prefers cold climates.
  - **Porcelain Garlic:** This variety produces large, plump cloves with a bold flavor and a hint of spiciness. Its bulbs have a beautiful, satiny sheen on their skin.

- ✠ **Purple Stripe Garlic:** Characterized by its vividly colored bulbs and striped, often mottled cloves. It has a strong, robust flavor.

Elephant Garlic (*Allium ampeloprasum*):

- ★ Despite its name, elephant garlic is not a true garlic but belongs to the leek family. It has large, mild-flavored cloves.
- ★ The bulbs are much larger than regular garlic, resembling a small onion.
- ★ Elephant garlic is often used as a milder substitute for true garlic in dishes where a subtler garlic flavor is desired.

Creole Garlic (Various Subspecies):

- ★ Creole garlic varieties are often associated with warm climates and are known for their rich and spicy flavors.
- ★ Varieties within the Creole group include:
  - ✠ **Ajo Rojo:** A Creole variety with reddish-brown skin and a strong, hot flavor. It's well-suited to hot climates.
  - ✠ **Inchelium Red:** A softneck Creole garlic with large cloves and a mild, slightly sweet flavor. It stores well and has a long shelf life.

Asian Garlic (*Allium sativum* var. *asiaticum*):

- ★ Asian garlic varieties are commonly grown in Asian countries and are known for their pungent and bold flavors.
- ★ Varieties within the Asian group include:
  - ✠ **Asian Tempest:** A fiery variety with a strong and spicy taste. It has small to medium-sized cloves and is ideal for those who enjoy intense garlic flavor.
  - ✠ **Japanese Garlic:** This variety produces bulbs with a complex, nutty flavor. The cloves are usually small and numerous.

Preparing Cloves for Planting:

- ★ Separate garlic bulbs into individual cloves just before planting.
- ★ Use larger, healthy cloves for planting, as they will produce larger bulbs.

Planting:

- ★ Plant garlic cloves with the pointed end up and the basal plate (root end) down.
- ★ Space the cloves about 6-8 inches apart in rows with a row spacing of around 12-18 inches.

- ★ Plant the cloves at a depth of 2-3 inches, covering them with soil.

Mulching:

- ★ Apply a layer of mulch (straw or chopped leaves) to protect the garlic from temperature fluctuations and suppress weed growth.

Fertilization:

- ★ Conduct a soil test to determine nutrient needs.
- ★ Apply a balanced fertilizer or compost in the fall before planting.

Watering:

- ★ Garlic requires consistent moisture, especially during the growing season.
- ★ Water the plants evenly to prevent under-watering or over-watering.

Weed Control:

- ★ Keep the garlic bed weed-free to prevent competition for nutrients and space.
- ★ Mulching also helps suppress weeds.

Scaping (for Hardneck Varieties):

- ★ Hardneck garlic varieties produce scapes, curly flower stalks.
- ★ Remove the scapes to divert the plant's energy into bulb growth.

Harvesting:

- ★ Garlic is ready to harvest when the lower leaves turn brown and dry out.
- ★ Dig carefully to avoid damaging the bulbs.
- ★ Allow the bulbs to cure in a dry, well-ventilated place with good airflow for about 2-3 weeks.

Storage:

- ★ After curing, brush off excess dirt and trim the roots and tops.
- ★ Store garlic bulbs in a cool, dry place with good ventilation to prevent mold and sprouting.

Saving Seed:

- ★ Select the best bulbs from your harvest for replanting next year.
- ★ Save the largest and healthiest cloves for planting to ensure consistent quality.

Potato

The potato (*Solanum tuberosum*) holds remarkable significance in agriculture, global food security, nutrition, and cultural heritage, making it one of the world's most important and widely consumed crops. Its multifaceted contributions extend from its role in addressing hunger to its adaptability in diverse climates, underscoring its importance in cultivation.

The potato stands as a vital source of sustenance for millions worldwide, particularly in regions where staple crops may be limited. Its energy-rich carbohydrates provide a staple food that can be prepared in various ways, helping address food scarcity and malnutrition. Rich in dietary fiber, vitamins (like vitamin C and B6), and minerals (like potassium), the potato contributes to balanced diets and better health outcomes, particularly in communities lacking diverse food options.One of the potato's remarkable qualities lies in its adaptability to diverse climates and terrains. It can thrive in various altitudes, from mountainous regions to lowlands, making it accessible to both rural and urban populations. Furthermore, the potato's culinary versatility enables it to take on various forms, from boiled and mashed to fried and baked, offering endless culinary possibilities and catering to diverse palates. Potatoes are renowned for their high yield per unit of land and relatively short growth cycle. This aspect makes them an attractive crop for small-scale farmers and communities striving for self-sufficiency. In subsistence farming, the potato's ability to provide a significant yield on limited land offers a buffer against food insecurity and economic instability.

The potato industry contributes significantly to economies globally, creating employment opportunities from production to processing, distribution, and marketing. This economic vitality supports rural livelihoods and contributes to both national and international trade.

Kufri Chandramukhi:

- **Origin:** Developed in India, particularly suitable for Indian climates.
- **Description:** Round to oval with light brown skin and white flesh. Moderately starchy.
- **Purpose:** Used for cooking a variety of Indian dishes, from curries to fries.

Kufri Jyoti:

- **Origin:** Developed in India for rain-fed conditions.

- **Description:** Round, with light pink skin and pinkish-white flesh. Waxy texture.
- **Purpose:** Ideal for boiling, salads, and as a side dish.

Kufri Sindhuri:

- **Origin:** Bred in India for table consumption.
- **Description:** Oval with bright red skin and light yellow flesh. Moderately starchy.
- **Purpose:** Suited for boiling, roasting, and mashing.

Atlantic:

- **Origin:** Developed in Canada.
- **Description:** Large, oval with light brown skin and white flesh. High starch content.
- **Purpose:** Commonly used for making French fries due to its high starch content.

Russet Burbank:

- **Origin:** Originated in the USA.
- **Description:** Large, elongated with russeted skin and white flesh. High starch content.
- **Purpose:** Preferred for French fries and baking due to its texture.

Yukon Gold:

- **Origin:** Developed in Canada.
- **Description:** Round to oval with light yellow skin and flesh. Medium starch.
- **Purpose:** Versatile for mashing, boiling, roasting, and frying.

Red Pontiac:

- **Origin:** Developed in the USA.
- **Description:** Round with red skin and white flesh. Waxy texture.
- **Purpose:** Well-suited for boiling, salads, and boiling as a whole.

Charlotte:

- **Origin:** Developed in France.

- ★ **Description:** Oval with pale yellow skin and yellow flesh. Waxy texture.
- ★ **Purpose:** Popular for boiling, salads, and as a new potato.

Maris Piper:

- ★ **Origin:** Originated in the UK.
- ★ **Description:** Oval with pale yellow skin and white flesh. Medium starch.
- ★ **Purpose:** Widely used for baking, frying, and making chips.

Purple Majesty:

- ★ **Origin:** Developed in the USA.
- ★ **Description:** Oblong with dark purple skin and flesh. High anthocyanin content.
- ★ **Purpose:** Adds vibrant color to dishes, suitable for roasting, and mashing.

Site Selection:

- ★ Choose a well-drained field with loose, loamy soil. Sandy or sandy-loam soils are preferable.
- ★ Ensure the selected site receives ample sunlight, ideally 6-8 hours per day.

Variety Selection:

- ★ Choose potato varieties suitable for your climate, soil type, and intended use (fresh consumption, processing, or storage).
- ★ Common varieties include Russet, Yukon Gold, Red Pontiac, and Kennebec.

Seed Preparation:

- ★ Purchase certified disease-free potato seed tubers from a reputable source.
- ★ Cut larger seed potatoes into pieces with at least one or two "eyes" (buds) per piece. Allow cut surfaces to air dry for a day or two before planting.

Planting:

- ★ Plant potatoes in the spring, a few weeks before the last expected frost date.

- ★ Dig trenches or furrows about 4-6 inches deep and place seed pieces with eyes facing up at a spacing of 12-18 inches apart.
- ★ Cover the seed pieces with soil and create small mounds or ridges above the planting rows.

Fertilization:

- ★ Conduct a soil test to determine nutrient requirements.
- ★ Apply a balanced fertilizer with higher phosphorus and potassium content, as excessive nitrogen can lead to excessive foliage growth.

Hilling or Earthing Up:

- ★ As the potato plants grow and reach about 6-8 inches tall, start hilling or earthing up the soil around the base of the plants.
- ★ This helps prevent the tubers from exposure to sunlight, which can cause them to turn green and become toxic.

Watering:

- ★ Potatoes require consistent moisture, especially during the tuber formation stage.
- ★ Avoid overwatering, which can lead to rot, and underwatering, which can result in smaller tubers.

Weed Control:

- ★ Keep the potato beds free from weeds, as they compete with the plants for nutrients and water.
- ★ Hand-weeding or using mulch can help control weed growth.

Pest and Disease Management:

- ★ Monitor for pests like Colorado potato beetles and aphids, and apply appropriate insecticides if necessary.
- ★ Watch for diseases such as late blight and early blight, and consider using disease-resistant varieties.

Harvesting:

- ★ Harvest time varies based on the potato variety and growing conditions, but it typically occurs when the plants start to yellow and die back.
- ★ Carefully dig up the tubers using a fork or shovel. Avoid damaging the tubers during the process.

Curing and Storage:

- ✯ Allow harvested potatoes to air dry and cure for a week in a cool, dry place with good ventilation.
- ✯ Store cured potatoes in a dark, cool (40-50°F or 4-10°C), and humid environment to prevent sprouting and maintain quality.

Tomato

The tomato (*Solanum lycopersicum*) stands as a cornerstone of global agriculture and culinary culture, its significance resonating through its nutritional value, culinary versatility, and agricultural impact. In both developing and developed nations, the tomato occupies a pivotal place in diets, contributing essential nutrients and vibrant flavor profiles to a myriad of dishes.

Nutritionally, tomatoes are a rich source of vitamins, especially vitamin C and vitamin A in the form of beta-carotene. They also offer significant amounts of antioxidants, including lycopene, which has garnered attention for its potential health benefits, particularly in reducing the risk of certain chronic diseases. This nutrient density supports overall health and underscores the tomato's role as a dietary asset.

Culturally, tomatoes have permeated the culinary tapestry of countless cultures, from Mediterranean to Asian cuisines. Their adaptability allows them to be utilized in diverse forms: raw in salads, cooked in sauces, roasted in entrees, or sun-dried for condiments. This versatility not only enhances gastronomic experiences but also lends itself to addressing culinary preferences and dietary requirements.

Agriculturally, tomatoes are a mainstay in commercial and subsistence farming systems alike. Their relatively short growing cycle and high yield potential contribute to food security, offering a reliable source of nutrition even in regions with shorter growing seasons. Additionally, tomatoes contribute to crop rotation strategies, helping manage soil-borne diseases and pests that affect other crops, thus supporting sustainable agricultural practices.

In economic terms, the global tomato industry is substantial, generating employment for farmers, processors, distributors, and retailers. It's a trade commodity that stimulates local economies and international markets, both in fresh and processed forms, including canned tomatoes, sauces, and pastes.

Furthermore, the tomato holds a special place in scientific research and breeding efforts, as it has been a model organism for studies on plant biology and genetics. The breakthroughs achieved through tomato research have had implications beyond the tomato itself, impacting the broader field of agricultural science.

In conclusion, the tomato's significance extends beyond its culinary appeal. It embodies nutritional fortitude, cultural enrichment, and agricultural sustainability. From nourishing millions to shaping agricultural practices and culinary artistry, the tomato's presence in our lives is testament to its enduring importance in the modern world.

Roma (Plum) Tomatoes:

- **Description:** Roma tomatoes are oval or cylindrical with fewer seeds and thicker flesh. They have a meaty texture.
- **Purpose:** Ideal for making sauces, pastes, and canning due to their low moisture content and rich flavor.

Cherry Tomatoes:

- **Description:** Cherry tomatoes are small, round, and usually bite-sized. They come in various colors, including red, yellow, and orange.
- **Purpose:** Often used in salads, appetizers, and snacking due to their sweet and refreshing taste.

Beefsteak Tomatoes:

- **Description:** Beefsteak tomatoes are large, juicy, and have a distinct ribbed shape. They come in different colors.
- **Purpose:** Perfect for slicing and adding to sandwiches, burgers, and salads, due to their juicy and substantial flesh.

Heirloom Tomatoes:

- **Description:** Heirloom tomatoes encompass a variety of non-commercialized, open-pollinated cultivars with unique flavors, colors, and shapes.
- **Purpose:** Appreciated for their diverse tastes and appearances, used in various culinary applications.

Campari Tomatoes:

- **Description:** Campari tomatoes are small and round with a deep red color and a sweeter taste.
- **Purpose:** Often used in salads, on pizzas, and as a snack due to their balanced sweetness and size.

Grape Tomatoes:

- ✯ **Description:** Grape tomatoes are small and elongated, resembling grapes. They have a firm texture and are typically sweeter.
- ✯ **Purpose:** Used in salads, garnishes, and as a flavorful addition to various dishes.

San Marzano Tomatoes:

- ✯ **Description:** San Marzano tomatoes are elongated with a pointed tip and thicker flesh. They are often canned.
- ✯ **Purpose:** Preferred for making sauces, as their dense flesh and fewer seeds create a rich flavor.

Green Zebra Tomatoes:

- ✯ **Description:** Green Zebra tomatoes are small, green, and have distinctive dark green stripes.
- ✯ **Purpose:** Used for their unique appearance and tangy flavor in salads, salsas, and garnishes.

Celebrity Tomatoes:

- ✯ **Description:** Celebrity tomatoes are round and medium-sized with a classic tomato flavor.
- ✯ **Purpose:** Versatile for various culinary applications, including sandwiches, salads, and sauces.

Yellow Pear Tomatoes:

- ✯ **Description:** Yellow pear tomatoes are small and pear-shaped, with a bright yellow color.
- ✯ **Purpose:** Often used for their vibrant color and sweet taste in salads, appetizers, and garnishes.

Cultivating tomatoes involves a series of steps to ensure healthy growth, proper fruit formation, and a successful harvest. Here's a general overview of the tomato cultivation process:

Site Selection:

- ✯ Choose a sunny location with at least 6-8 hours of direct sunlight daily.
- ✯ Ensure well-draining soil with a slightly acidic to neutral pH (around 6.0-7.0).

Seedling Preparation:

- Start tomato seeds indoors 6-8 weeks before the last expected frost date in your area.
- Use high-quality seed starting mix and provide adequate light and warmth for germination.

Transplanting:

- Transplant tomato seedlings outdoors after the danger of frost has passed and the soil has warmed up.
- Space plants about 2-3 feet apart in rows or raised beds.

Support and Trellising:

- Tomatoes benefit from support to prevent the plants from sprawling on the ground.
- Stake or cage the plants to keep them upright and to promote better air circulation, reducing disease risk.

Fertilization:

- Conduct a soil test to determine nutrient needs.
- Apply a balanced fertilizer or one with a higher phosphorus content to encourage flowering and fruiting.

Watering:

- Provide consistent moisture to the plants, particularly during flowering and fruiting.
- Water at the base of the plants to prevent fungal diseases on the leaves.

Mulching:

- Apply organic mulch (such as straw or compost) around the base of the plants to retain soil moisture, suppress weeds, and maintain even soil temperature.

Pruning:

- Remove the suckers that form in the leaf axils (crotches) of tomato plants to encourage better fruit production.
- Some pruning techniques include determinate and indeterminate types, so research the pruning method that suits your variety.

Pest and Disease Management:

- ✯ Regularly inspect plants for pests like aphids, hornworms, and whiteflies. Apply appropriate treatments if necessary.
- ✯ Watch for diseases like early and late blight, and use disease-resistant varieties or apply fungicides when needed.

Fruit Harvesting:

- ✯ Harvest tomatoes when they are fully ripe but still firm.
- ✯ Gently twist or cut the stem just above the fruit when harvesting.

Post-Harvest Handling:

- ✯ Handle tomatoes carefully to avoid bruising.
- ✯ Store ripe tomatoes at room temperature and use within a few days. Avoid storing them in the refrigerator, as it can affect flavor and texture.

Saving Seeds (Optional):

- ✯ If desired, you can save seeds from open-pollinated or heirloom tomato varieties for future planting.

Cucurbitaceae

Cucurbits, encompassing a diverse family of plants including cucumbers, squash, pumpkins, and melons, hold profound significance in agriculture, nutrition, and culture due to their wide-ranging culinary uses, nutritional value, economic impact, and contributions to sustainable farming practices.

- ✯ **Nutritional Value and Culinary Versatility:** Cucurbits offer an array of essential nutrients such as vitamins A and C, potassium, and dietary fiber. Their significance lies in their ability to enhance the nutritional content of diets. From refreshing salads made with cucumbers to hearty stews enriched with squash and the sweet indulgence of melons, cucurbits cater to a broad spectrum of culinary preferences. Their versatility extends to various forms of consumption, from raw and fresh to cooked and preserved, making them staples in diverse cuisines worldwide.
- ✯ **Economic Impact and Agriculture:** The cultivation of cucurbits has substantial economic implications, providing livelihoods for countless farmers, processors, distributors, and retailers. These crops contribute to

local and global agricultural economies through direct consumption and processed products. Their relatively short growing cycles make them feasible options for both small-scale and large-scale agricultural systems, ensuring a consistent supply of produce to markets. Additionally, cucurbits are often intercropped, which enhances biodiversity and supports sustainable agricultural practices.

- **Disease and Pest Management:** Cucurbits play a strategic role in integrated pest management and crop rotation strategies. Certain cucurbit crops can break disease cycles by reducing the build-up of soil-borne pathogens that affect other crops. By acting as a "trap crop," they draw pests away from main crops, thereby reducing the need for chemical interventions. This natural pest management approach contributes to reduced pesticide usage and more environmentally friendly farming practices.
- **Cultural Significance:** Cucurbits have deep-rooted cultural significance in many societies. From traditional recipes passed down through generations to festive celebrations featuring carved pumpkins, these plants weave themselves into the fabric of cultural practices. In some cultures, cucurbit seeds and oils hold symbolic and medicinal importance. The integration of cucurbits into traditional cuisines reflects their role in sustaining cultural identities.
- **Sustainability and Biodiversity:** Cucurbits contribute to crop diversity and resilience in farming systems. Their incorporation in crop rotations and agroforestry helps maintain soil health and structure, reducing the need for excessive chemical inputs. Furthermore, heirloom and indigenous cucurbit varieties contribute to the conservation of genetic diversity, safeguarding plants adapted to specific ecological niches and changing environmental conditions.

Cucurbits encompass a diverse group of plants that includes cucumbers, squash, pumpkins, and melons. These plants are cultivated for their edible fruits and are popular additions to gardens and agricultural systems.

Here's a comprehensive guide to cultivating cucurbits, along with descriptions of some common varieties:

Site Selection:

- Choose a sunny location with 6-8 hours of direct sunlight per day.
- Opt for well-drained soil with good fertility and a pH level of 6.0-7.0.

Seedling Preparation:

- ✯ Start seeds indoors a few weeks before the last expected frost date in your area.
- ✯ Use a well-draining seed starting mix and provide warmth and light for germination.

Transplanting:

- ✯ Transplant seedlings outdoors when all danger of frost has passed and the soil has warmed up.
- ✯ Space plants according to their growth habits and recommended spacing.

Support and Trellising:

- ✯ Some cucurbits benefit from vertical support to save space and prevent disease.
- ✯ Install trellises or cages for plants like cucumbers and certain squash varieties.

Fertilization:

- ✯ Conduct a soil test to determine nutrient requirements.
- ✯ Apply a balanced fertilizer or one higher in phosphorus to promote flowering and fruiting.

Watering:

- ✯ Maintain consistent soil moisture, especially during flowering and fruiting stages.
- ✯ Water at the base of the plants to prevent fungal diseases on the leaves.

Mulching:

- ✯ Apply organic mulch around the plants to retain moisture, suppress weeds, and maintain even soil temperature.

Pruning and Training:

- ✯ Some cucurbits benefit from limited pruning to direct energy to fruit production and improve air circulation.

Pest and Disease Management:

- ✯ Regularly inspect plants for pests like aphids, squash bugs, and cucumber beetles.

- ★ Monitor for diseases like powdery mildew and downy mildew and use disease-resistant varieties or apply fungicides when needed.

Fruit Harvesting:

- ★ Harvest fruits when they are mature but before they become overripe.
- ★ Use sharp tools to cut fruits from the vine to prevent damage.

Common Cucurbit Varieties and Descriptions:

Cucumbers (*Cucumis sativus*):

- ★ Varieties: 'Marketmore,' 'Straight Eight,' 'Lemon,' 'Armenian'
- ★ Descriptions: Cucumbers are typically green and elongated, but some varieties are yellow or curved. They are often used fresh in salads or pickled.

Zucchini and Summer Squash (*Cucurbita pepo*):

- ★ Varieties: 'Black Beauty,' 'Yellow Crookneck,' 'Pattypan,' 'Costata Romanesco'
- ★ Descriptions: Zucchini and summer squash come in various shapes and colors. They are versatile for cooking and can be used in stir-fries, grilling, and baking.

Winter Squash (*Cucurbita maxima* and *Cucurbita moschata*):

- ★ Varieties: 'Butternut,' 'Acorn,' 'Spaghetti,' 'Hubbard'
- ★ Descriptions: Winter squash have hard shells and include a range of flavors and textures. They're often used for roasting, soups, and baking.

Pumpkins (*Cucurbita pepo*):

- ★ Varieties: 'Sugar Pie,' 'Jack O' Lantern,' 'Cinderella,' 'Jarrahdale'
- ★ Descriptions: Pumpkins vary in size and color. Sugar Pie pumpkins are known for their sweet flesh, while others are often carved for Halloween.

Melons (*Cucumis melo*):

- ★ Varieties: 'Cantaloupe,' 'Honeydew,' 'Watermelon,' 'Muskmelon'
- ★ Descriptions: Melons have sweet and aromatic flesh. They're enjoyed fresh and are popular summer fruits.

Gourds (*Lagenaria siceraria*):

- ★ Varieties: 'Bottle Gourd,' 'Luffa Gourd,' 'Birdhouse Gourd'

- ★ Descriptions: Gourds come in various shapes and sizes. Some are edible when young, while others are used for crafting or decoration.

Cultivating cucurbits offers a wide range of flavorful and nutritious fruits for various culinary applications. While specific care varies among cucurbit types, attention to water, nutrients.

6

Plant Protection Strategies for Vegetable Crops

Ensuring the health and vitality of vegetable crops is a critical aspect of successful agriculture. In this chapter, we explore the multifaceted world of plant protection strategies for vegetable crops. From understanding pest life cycles to implementing integrated pest management, we delve into the techniques and practices that safeguard the quality and yield of these valuable plants.

The Challenge of Pest Management

Vegetable crops are vulnerable to a wide range of pests, including insects, pathogens, and weeds, which can cause substantial yield losses if left uncontrolled. The delicate balance between promoting plant growth and managing pest populations requires a holistic approach to pest management. Chemical pesticides have historically been a primary tool for controlling pests. However, concerns about environmental impact, resistance, and human health have prompted the exploration of alternative and sustainable strategies.

Integrated Pest Management (IPM)

Integrated Pest Management (IPM) is a holistic and sustainable approach that aims to manage pests by combining multiple strategies. IPM integrates biological, cultural, physical, and chemical measures to control pests while minimizing adverse effects on human health, non-target organisms, and the environment. One of the cornerstones of IPM is biological control, where natural predators, parasites, and pathogens are introduced to regulate pest populations. This approach reduces the reliance on chemical pesticides and encourages natural ecological balance. Cultural practices also play a pivotal role in IPM. Crop rotation, intercropping, and companion planting are techniques that disrupt pest life cycles, reduce pest populations, and enhance soil health.

Resistant Varieties and Biotechnology

Breeding for resistance is a proactive approach to pest management. Plant breeders develop resistant varieties that are less susceptible to pests and diseases. This not only reduces the need for chemical controls but also prolongs the efficacy of existing management strategies.

Biotechnology, including genetic engineering, offers innovative solutions in plant protection. Genetically modified (GM) crops can express traits that confer resistance to specific pests or diseases, reducing the need for external interventions.

Chemical and Alternative Control Methods

While moving towards sustainable practices is encouraged, chemical control methods still play a role in plant protection. However, the emphasis is on judicious and targeted use to minimize environmental impact and reduce the development of resistance. Alternative methods, such as insecticidal soaps, oils, and botanical extracts, offer more environmentally friendly options for managing pests. These substances have minimal residual effects and are often specific to certain pest groups.

Physical control measures, like barriers, traps, and reflective mulches, can disrupt the behavior and movement of pests. These techniques are particularly valuable in organic farming systems.

In this chapter, we've explored the intricate world of plant protection strategies for vegetable crops. From the challenges of pest management to the holistic approach of Integrated Pest Management (IPM), we've journeyed through methods that balance productivity with environmental stewardship. The quest for sustainable solutions continues as we navigate through the complexities of pest life cycles, the art of biological control, and the role of modern biotechnology in shaping the future of plant protection. As we move forward, we'll uncover post-harvest considerations and explore the dynamic relationship between vegetables and human health.

7

Vegetable Breeding: Nurturing Nature's Bounty

Principles of Plant Breeding

Plant breeding is both an art and a science that has been practiced for centuries, with the goal of enhancing the qualities of crops to meet the evolving needs of human societies. Vegetable breeding specifically focuses on improving the characteristics of plants that are cultivated for human consumption, including factors such as yield, nutritional content, disease resistance, flavor, and appearance. At its core, plant breeding is guided by several fundamental principles:

Genetic Variation

Genetic variation is the foundation of plant breeding. It provides the raw material from which breeders can select and develop new vegetable varieties. By identifying and harnessing natural variations in traits, breeders can create plants that are better suited to changing environmental conditions and consumer preferences.

Selection

Selection involves identifying plants with desirable traits and using them as parents for the next generation. This process is repeated over multiple generations to accumulate the desired traits in the offspring. Selection can be done based on observable traits, as well as through molecular markers that are associated with specific traits.

Hybridization

Hybridization is the controlled crossing of two different plants to create offspring with a combination of desirable traits from both parents. It often results in "hybrid vigor" or increased growth and performance in the offspring. Hybrid vegetable varieties are widely cultivated due to their improved yield, uniformity, and other advantageous qualities.

Segregation

In subsequent generations, the traits of hybrids and their offspring segregate due to the inheritance of different combinations of genes. This provides opportunities for breeders to select and stabilize specific traits through further rounds of selection.

Breeding Techniques: Conventional and Modern Approaches

Conventional Breeding

Conventional breeding techniques have been employed for centuries and involve selecting and crossbreeding plants based on observable traits. This method often relies on the expertise of breeders in identifying plants with desired characteristics and making controlled crosses. The main limitation of conventional breeding is the time it takes to accumulate and stabilize traits over generations.

Modern Approaches

Modern breeding techniques, including biotechnology and molecular genetics, have revolutionized the field of vegetable breeding. These approaches allow breeders to identify and manipulate specific genes associated with desired traits. For example, genetic markers can be used to track the presence of target genes throughout generations, making selection more precise and efficient. Genetic modification techniques like CRISPR-Cas9 have enabled the direct editing of specific genes, accelerating the development of new varieties with targeted improvements.

Hybridization and Selection for Desired Traits

Hybridization Techniques

Hybrid vegetable varieties are widely cultivated due to their increased vigor and uniformity. To produce hybrids, breeders carefully select parent plants with complementary traits and perform controlled crosses. The resulting hybrid offspring often exhibit a phenomenon known as hybrid vigor or heterosis, which can lead to improved yield, disease resistance, and other desirable qualities.

Selection for Desired Traits

Selection is a crucial step in vegetable breeding, allowing breeders to choose plants with the most desirable characteristics to serve as parents for the next generation. Traits such as yield, quality, disease resistance, and adaptability to specific growing conditions are carefully assessed and selected. Modern molecular techniques, including marker-assisted selection (MAS) and genomic selection, enable breeders to identify plants with the desired genes more efficiently.

Role of Genetics in Improving Vegetable Traits

Introduction

Vegetable crops play a vital role in global food security and human nutrition. The continuous growth of the global population poses challenges to meet the rising demand for nutritious and diverse vegetables. Genetic improvement of vegetable crops has been a cornerstone in enhancing crop productivity, quality, and resilience to environmental stresses. This chapter delves into the pivotal role of genetics in improving vegetable traits, highlighting key genetic techniques, their applications, and significant advancements in the field.

Classical Breeding and Genetic Variation

Classical breeding has been the foundation of improving vegetable traits for centuries. Breeders select plants with desirable traits, such as higher yield, disease resistance, and improved taste, and then cross them to create new hybrids with improved characteristics. This method heavily relies on genetic variation within and among populations. Genetic diversity is a critical factor in creating a reservoir of genes that can be tapped into for various traits.

Marker-Assisted Selection (MAS)

Advancements in molecular genetics have revolutionized the way vegetable traits are improved. Marker-assisted selection (MAS) is a technique that uses molecular markers linked to specific traits of interest to aid in traditional breeding. This technique accelerates the breeding process by allowing breeders to identify and select plants with desired traits at an early stage, without having to wait for the traits to become visible.

Genetic Engineering and Transgenics

Genetic engineering has enabled the direct manipulation of an organism's DNA, leading to the creation of genetically modified (GM) crops. In the realm of vegetables, genetic engineering has been used to confer traits such as insect resistance, herbicide tolerance, and improved shelf life. However, the adoption of GM vegetables has been met with both enthusiasm and skepticism due to concerns related to environmental impacts and food safety.

Genome Editing and CRISPR-Cas9

The development of CRISPR-Cas9 technology has ushered in a new era of precision genome editing. This tool allows scientists to make targeted changes to an organism's DNA, including vegetables, with unprecedented accuracy. Genome editing has the potential to improve traits like flavor, nutritional content, and disease resistance without introducing foreign DNA, addressing some of the concerns associated with traditional genetic engineering.

Omics Technologies and Trait Mapping

Omics technologies, including genomics, transcriptomics, proteomics, and metabolomics, have provided insights into the molecular basis of vegetable traits. By analyzing the complete set of genes, proteins, and metabolites, researchers can identify key genes associated with specific traits. This information is essential for accelerating breeding efforts by guiding the selection of plants with desirable genetic markers.

8

Vertical Gardening

Vertical gardening is a space-saving and innovative technique that involves growing vegetables vertically on structures like walls, trellises, fences, and containers. This approach maximizes your growing area and allows you to cultivate a variety of vegetables even in limited spaces. Here are details on vertical gardening with vegetables:

1. Selection of Vegetables:

Choose vegetables that naturally climb, have vining habits, or can be trained to grow vertically. Some suitable options include:

- **Tomatoes:** Cherry or grape tomatoes work well in vertical setups.
- **Cucumbers:** Choose compact or bush cucumber varieties.
- **Beans:** Pole beans or runner beans are excellent climbers.
- **Peas:** Select dwarf or climbing pea varieties.
- **Squash:** Compact zucchini or scallop squash can be grown vertically.

2. Vertical Structures:

- Install trellises, arbors, stakes, cages, or even repurpose pallets as vertical supports.
- Determine the height and sturdiness of the structure based on the vegetable's growth habit and weight.

3. Planting:

- Plant seeds or seedlings at the base of the structure, following spacing recommendations on seed packets or plant labels.
- Ensure proper soil preparation and use well-draining soil with good organic content.

4. Training and Pruning:

- ✯ Gently guide the plants to grow along the vertical support as they climb.
- ✯ Remove any side shoots or suckers to encourage upward growth and redirect energy to fruit production.

5. Support Systems:

- ✯ Use soft ties, garden twine, clips, or plant clips to secure the plants to the vertical support.
- ✯ As the plants grow, you may need to adjust the ties to prevent constriction.

6. Watering and Fertilization:

- ✯ Maintain consistent moisture in the soil, especially in containers or raised beds.
- ✯ Fertilize with balanced organic fertilizers or compost to provide essential nutrients for vigorous growth.

7. Pollination:

- ✯ Some vertically grown vegetables, like cucumbers and squash, may require pollination for fruit set. Hand-pollination can be done if pollinators are scarce.

8. Harvesting:

- ✯ Harvest vegetables as they mature. Regular harvesting encourages continuous production.
- ✯ Be careful while harvesting not to damage the plants or the structure.

9. Vertical Gardening Ideas:

- ✯ Use hanging baskets or vertical pockets to grow herbs, lettuce, and smaller vegetables.
- ✯ Attach planter boxes to walls or fences for a cascading effect.
- ✯ Create a living wall with a variety of edible plants.
- ✯ Build a pyramid-shaped trellis for multi-level growing.

10. Considerations:

- ✯ Ensure that the chosen structure can support the weight of the plants and the developing fruits.

- ★ Adequate sunlight is crucial. Position the vertical garden where it receives sufficient sunlight for healthy growth.

Vertical gardening with vegetables not only saves space but also adds an aesthetic element to your garden. With proper care, you can enjoy a productive and visually appealing garden that makes the most of your available space.

Storage of Vegetable Crops

Introduction

Storage is a critical phase in the post-harvest handling of vegetable crops, aimed at preserving the quality and extending the shelf life of produce. Proper storage practices are essential to minimize losses due to spoilage, decay, and physiological deterioration. This chapter explores various aspects of vegetable crop storage, including factors affecting storage, types of storage facilities, and techniques to maintain optimal storage conditions.

Factors Affecting Vegetable Crop Storage

Successful vegetable crop storage depends on understanding and managing several factors that influence post-harvest quality:

Temperature

Temperature control is a key factor in vegetable storage. Each vegetable has an optimal temperature range that slows down metabolic processes and reduces spoilage. For instance, root vegetables like carrots and potatoes are typically stored at cooler temperatures (32°F to 50°F or 0°C to 10°C), while tomatoes and peppers require slightly higher temperatures (50°F to 60°F or 10°C to 15°C).

Humidity

Maintaining proper humidity levels helps prevent dehydration and wilting. Leafy greens and some cruciferous vegetables prefer higher humidity levels, while low humidity is more suitable for root crops. Condensation control is crucial to avoid mold and bacterial growth.

Ethylene Sensitivity

Ethylene is a natural plant hormone that can accelerate ripening and spoilage in ethylene-sensitive vegetables like lettuce, broccoli, and cucumbers. Proper storage should separate ethylene-producing crops (e.g., tomatoes, bananas) from ethylene-sensitive ones.

Ventilation

Adequate air exchange helps remove heat, humidity, and gases produced by respiration. Lack of ventilation can lead to the accumulation of carbon dioxide and ethylene, accelerating spoilage. Ventilation requirements vary depending on the type of vegetable.

Light

Exposure to light can cause chlorophyll degradation and off-flavors in certain vegetables. Potatoes and onions, for example, should be stored in darkness to prevent greening.

Types of Storage Facilities

Several types of storage facilities are utilized for vegetable crops, each designed to meet specific requirements:

Cold Storage

Cold storage facilities include walk-in coolers and refrigerated warehouses. These facilities maintain low temperatures and controlled humidity levels. They are suitable for a wide range of vegetables, providing extended shelf life while preserving quality.

Controlled Atmosphere (CA) Storage

CA storage involves modifying the atmosphere within storage rooms to control oxygen, carbon dioxide, and ethylene levels. This technique can significantly extend the shelf life of vegetables by reducing respiration rates and inhibiting microbial growth.

Root Cellars

Root cellars are traditional storage structures dug into the ground, providing natural temperature and humidity regulation. They are particularly suitable for root vegetables and certain fruits, maintaining conditions favorable for extended storage.

Ventilated Storage

Ventilated storage structures, such as mesh bags or crates with gaps, facilitate air exchange. These are useful for crops like onions and garlic, allowing proper ventilation to prevent mold and spoilage.

Storage Techniques

Implementing effective storage techniques is crucial for maintaining post-harvest quality:

Pre-cooling

Rapidly cooling vegetables after harvest minimizes deterioration. Hydrocooling, vacuum cooling, and forced-air cooling are common pre-cooling methods that reduce field heat and remove moisture.

Packaging

Appropriate packaging can protect vegetables from physical damage, reduce water loss, and control gas exchange. Perforated plastic bags, modified atmosphere packaging (MAP), and vacuum-sealing are packaging methods that extend shelf life.

Monitoring and Management

Regular monitoring of temperature, humidity, and ethylene levels is essential for preventing spoilage. Technological advancements allow for remote monitoring and alerts, enabling timely corrective actions.

Proper storage practices are instrumental in maintaining the freshness and quality of vegetable crops. By understanding the factors influencing storage, utilizing appropriate storage facilities, and implementing effective storage techniques, post-harvest losses can be minimized, and consumers can enjoy nutrient-rich vegetables with extended shelf life.

9
Marketing Strategies for Fresh Vegetables

Introduction

In today's dynamic market, the marketing of vegetables has evolved from a simple transactional process to a complex and multifaceted endeavor. The increased emphasis on health-conscious lifestyles and sustainable consumption has elevated the importance of fresh vegetables. This chapter delves into the strategies employed for marketing vegetables, considering both traditional methods and modern digital approaches. Drawing on established theories and real-world examples, this chapter aims to provide insights into the art of successfully marketing vegetables in a competitive landscape.

Understanding Consumer Behavior

Before delving into marketing strategies, it's essential to grasp consumer behavior when it comes to vegetable purchases. Various psychological and economic factors influence consumer choices. The "Theory of Planned Behavior" by Ajzen (1991) provides insights into how attitudes, subjective norms, and perceived behavioral control impact buying decisions. Marketers can tailor their strategies to address these factors and promote vegetable consumption.

Segmentation and Targeting

Segmentation involves dividing the market into distinct groups with shared characteristics. Through market segmentation, vegetable marketers can identify specific consumer needs and preferences. For instance, some consumers prioritize organic produce due to health concerns, while others seek affordability. By targeting these segments individually, marketers can develop more focused campaigns and product offerings.

Product Differentiation and Positioning

Vegetable marketers can employ product differentiation to set their offerings apart from competitors. This can involve emphasizing factors like freshness, sourcing practices, or unique varieties. Positioning, on the other hand, entails how a product is perceived in the minds of consumers. Effective positioning can leverage factors like quality, convenience, or health benefits.

Promotion and Communication Strategies

Promotion plays a pivotal role in marketing vegetables. Leveraging both traditional and digital channels, marketers can create awareness and drive sales. Social media platforms, for example, allow for engaging content creation and direct consumer interaction. Additionally, content marketing, such as recipe blogs and nutritional articles, can position the brand as a reliable source of information and recipes, fostering consumer trust.

Distribution Channels

Efficient distribution is crucial for ensuring vegetables reach consumers in optimal condition. Farmers' markets, grocery stores, online platforms, and even subscription services offer diverse distribution channels. The choice of channel often depends on the target market and the brand's logistical capabilities.

Sustainability and Ethical Marketing

With increasing awareness of environmental issues, sustainability has become a powerful marketing tool. By emphasizing sustainable farming practices, reduced carbon footprint, and ethical sourcing, vegetable marketers can attract environmentally conscious consumers. The "Green Marketing" concept (Peattie & Peattie, 2003) highlights the potential benefits of aligning marketing strategies with ecological responsibility.

Building Brand Loyalty

Long-term success in vegetable marketing hinges on brand loyalty. A positive customer experience, consistent quality, and personalized interactions contribute to strong brand relationships. Loyalty programs, personalized recommendations, and post-purchase engagement can all foster customer retention.

Case Study: "Farm to Table Organics"

As an illustration of effective vegetable marketing, consider the case of "Farm to Table Organics." This brand positioned itself as a local, sustainable, and health-conscious choice for consumers seeking premium organic vegetables. They used a combination of farmers' market presence, an online store, and educational blogs to engage with their audience. By transparently showcasing their farming practices and emphasizing the nutritional benefits of their produce, they successfully established a loyal customer base.

Conclusion

Marketing vegetables involves a blend of understanding consumer behavior, creating targeted strategies, and embracing sustainable practices. By utilizing segmentation, differentiation, effective communication, and sustainable initiatives, vegetable marketers can thrive in a competitive market. Keeping an eye on emerging trends and consumer preferences will continue to be crucial for adapting marketing approaches to the ever-evolving landscape of vegetable consumption.

Exporting Fresh Vegetables: Strategies and Challenges

Introduction

The global trade in fresh vegetables has experienced remarkable growth, driven by increased consumer demand for diverse produce year-round and advancements in transportation and logistics. Exporting vegetables presents both opportunities and challenges for producers, exporters, and importers. This chapter explores the strategies employed in the export of fresh vegetables, taking into consideration market dynamics, regulatory requirements, and logistical intricacies.

Understanding Global Market Dynamics

The export of fresh vegetables involves navigating diverse international markets with varying preferences, regulations, and cultural norms. Market research is pivotal for identifying potential export destinations, understanding consumer preferences, and gauging the competitive landscape. The use of Porter's Five Forces model can help assess market attractiveness and competitive intensity (Porter, 1980).

Quality Control and Compliance

Ensuring the quality and safety of exported vegetables is paramount to maintaining customer trust and complying with international regulations. Adherence to international standards such as those set by the Codex Alimentarius Commission ensures the safety and quality of food products traded globally (FAO/WHO, 2007). Implementing Good Agricultural Practices (GAP) and following guidelines for post-harvest handling can help preserve the freshness and nutritional value of vegetables during transportation.

Logistics and Supply Chain Management

The success of vegetable exports heavily relies on efficient logistics and supply chain management. The "cold chain," a temperature-controlled supply chain, is essential to prevent deterioration of perishable goods. Effective coordination among growers, packers, exporters, and importers is necessary to maintain product quality from farm to consumer. Modern technologies like blockchain can enhance transparency and traceability within the supply chain (Iansiti & Lakhani, 2020).

Trade Agreements and Tariffs

Bilateral and multilateral trade agreements play a significant role in facilitating vegetable exports. Reduction or elimination of tariffs through trade agreements can enhance a country's competitiveness in foreign markets. Additionally, understanding rules of origin is crucial to benefit from preferential tariff treatment under various trade pacts.

Market Entry Strategies

Exporters can adopt different market entry strategies, such as direct exports, joint ventures, or licensing agreements, depending on the target market's characteristics. Collaborating with local distributors or partners can help navigate cultural nuances, regulatory hurdles, and distribution challenges.

Case Study: The Netherlands' Horticultural Exports

The Netherlands is renowned for its successful export of fresh vegetables, particularly its horticultural products. The country's strategic investments in research and technology, along with its well-developed logistics and distribution networks, have positioned it as a major player in the global vegetable trade. The Dutch approach emphasizes innovation, sustainable practices, and a strong focus on quality control.

Challenges and Future Outlook

While exporting vegetables presents lucrative opportunities, challenges persist. These include fluctuating demand, strict phytosanitary regulations, geopolitical tensions, and sustainability concerns. The future of vegetable exports will likely be shaped by advancements in technology, increased emphasis on sustainable practices, and the continued evolution of global trade dynamics.

Exporting fresh vegetables requires a comprehensive understanding of market dynamics, quality control measures, logistics, and trade regulations. By embracing quality assurance, efficient supply chain management, and strategic market entry approaches, exporters can tap into the growing global demand for diverse and fresh vegetables.

Success Stories in Indian Vegetable Farming

Introduction:

India's agricultural landscape has undergone significant transformations over the years, with vegetable farming playing a pivotal role in the country's food security and economic growth. This chapter delves into some remarkable success stories in Indian vegetable farming, highlighting innovative practices, modern technologies, and sustainable approaches that have propelled farmers to achieve remarkable yields and profits. These case studies showcase the resilience and adaptability of Indian farmers, providing valuable insights for the global agricultural community.

Integrated Pest Management in Tomato Cultivation:

One notable success story is that of Rajesh Singh, a progressive farmer from Punjab, who transformed his tomato cultivation practices through integrated pest management (IPM). Facing recurring pest issues that threatened his yields, Singh adopted a holistic approach, combining natural predators, resistant varieties, and judicious use of biopesticides. This not only reduced chemical inputs but also increased his yields by 30% over two growing seasons .

Precision Irrigation in Onion Fields:

In the arid region of Rajasthan, Sunita Patel pioneered the use of precision irrigation techniques in her onion fields. By leveraging soil moisture sensors and weather data, she optimized irrigation scheduling, reducing water wastage by 40% while improving onion quality and size. This innovation not only conserved water but also enhanced her farm's profitability.

Organic Farming in the Western Ghats:

Moving to the Western Ghats, we find the story of Anand Kumar, an organic vegetable farmer who revived traditional farming practices. Focusing on soil health and biodiversity, Kumar implemented cover cropping, composting, and crop rotation. His farm not only achieved organic certification but also witnessed a revival of native flora and fauna, demonstrating the potential of sustainable agriculture.

Hydroponics in Urban Farming:

In the bustling city of Mumbai, Ramesh Gupta redefined urban farming by adopting hydroponics for growing lettuce and herbs. By utilizing vertical growing systems and nutrient-enriched solutions, Gupta produced high-quality vegetables in limited urban spaces, reducing the city's dependence on external vegetable supplies .

Community-Based Seed Banks in KeralA.

The state of Kerala embraced community-based seed banks to safeguard traditional vegetable varieties. Mary Mathew led a grassroots movement to establish these seed banks, preserving hundreds of native vegetable varieties and ensuring local food security. The initiative gained global recognition as an example of preserving agrobiodiversity.

Conclusion:

These success stories illuminate the diversity and ingenuity of Indian vegetable farmers. From technological innovations to sustainable practices, these farmers have not only improved their own livelihoods but have also contributed to India's agricultural progress. Their experiences provide valuable insights into addressing the challenges faced by vegetable farmers globally. As India continues to evolve agriculturally, these stories serve as beacons of inspiration for the rest of the world.

Ethical and Societal Considerations in Vegetable Farming

Introduction:

As vegetable farming continues to play a crucial role in global food systems, ethical and societal considerations are becoming increasingly important. This chapter explores the multifaceted dimensions of these considerations, including fair trade practices in vegetable trade, the social and economic impacts of vegetable farming, and the pivotal role of vegetables in ensuring food security and promoting human health. By analyzing these aspects, we gain insights into the broader implications of vegetable farming beyond mere agricultural production.

Fair Trade Practices in Vegetable Trade:

Fair trade practices have gained prominence as a means to ensure equitable and ethical relationships between producers and consumers in the global market. In the context of vegetable trade, organizations such as Fair trade International have established standards to guarantee fair wages, safe working conditions, and community development for small-scale vegetable farmers. Case studies from countries like Kenya and Peru showcase how fair trade initiatives empower farmers to improve their livelihoods and invest in sustainable farming practices.

Social and Economic Impacts of Vegetable Farming:

The social and economic impacts of vegetable farming extend beyond the farm gate, influencing entire communities and regions. In rural areas, vegetable cultivation provides employment opportunities, especially for women and marginalized groups, contributing to poverty reduction and gender equality. Moreover, vegetable farming diversifies income sources and enhances resilience against crop failures. Research in India, for instance, demonstrates that small holder vegetable farming contributes significantly to rural economies and household nutrition.

Role of Vegetables in Food Security and Health:

Vegetables are a cornerstone of food security and human health. They provide essential nutrients, vitamins, and antioxidants necessary for well-being and disease prevention. Vegetable farming contributes to a diversified diet, combating malnutrition and promoting long-term health. Initiatives like school vegetable gardens in South Africa have shown that involving children in vegetable production not only enhances their dietary habits but also imparts agricultural skills and environmental awareness.

Conclusion:

Ethical and societal considerations are integral to understanding the broader implications of vegetable farming. Fair trade practices ensure that the benefits of vegetable trade are distributed equitably among producers, while the social and economic impacts of vegetable farming go beyond economic gains, reaching into the realms of community development and gender equality. Furthermore, the role of vegetables in food security and health underscores their significance in achieving sustainable and resilient food systems. By embracing these considerations, stakeholders can collectively work towards a more just, nutritious, and sustainable vegetable farming industry.

Flowers

1

Introduction to Ornamental Plants

Ornamental plants, also known as decorative or flowering plants, are cultivated primarily for their aesthetic appeal and decorative qualities. They enhance the beauty of landscapes, gardens, indoor spaces, and public areas, adding color, texture, and visual interest. Ornamental plants encompass a wide range of species, from flowers and shrubs to trees and vines. Here's a detailed introduction to ornamental plants with examples:

Flowers:

- **Roses (*Rosa* spp.):** Known for their elegance and fragrance, roses come in various colors and types, including hybrid tea roses, floribundas, and climbers.
- **Tulips (*Tulipa* spp.):** These spring-blooming bulbs showcase a vibrant array of colors and shapes, making them a popular choice for gardens and landscapes.
- **Lilies (*Lilium* spp.):** Lilies are known for their large, showy blooms and come in various forms such as Asiatic, Oriental, and trumpet lilies.
- **Petunias (*Petunia* spp.):** Petunias are versatile annuals that produce colorful, trumpet-shaped flowers in various hues.

Shrubs and Bushes:

- **Azaleas (*Rhododendron* spp.):** These flowering shrubs are admired for their striking clusters of vibrant flowers and evergreen foliage.
- **Hydrangeas (*Hydrangea* spp.):** Hydrangeas are loved for their large, globe-like flower clusters that change color based on soil pH.

- ★ **Forsythias (*Forsythia* spp.):** Early-blooming shrubs with bright yellow flowers that signal the arrival of spring.
- ★ **Lilacs (*Syringa* spp.):** Lilacs are known for their fragrant, cone-shaped flower clusters and are often associated with nostalgia.

Trees:

- ★ **Cherry Blossoms (*Prunus* spp.):** Cherished for their delicate pink or white blossoms, cherry trees are often symbols of beauty and transience.
- ★ **Japanese Maple (*Acer palmatum*):** These trees are prized for their vibrant foliage in shades of red, orange, and purple.
- ★ **Magnolias (*Magnolia* spp.):** Magnolias are admired for their large, fragrant flowers that often bloom before the leaves emerge.

Vines and Climbers:

- ★ **Clematis (*Clematis spp.*):** These climbing plants produce an array of showy flowers in various colors and shapes.
- ★ **Wisteria (*Wisteria spp.*):** Known for their cascading clusters of fragrant flowers, wisterias create a dramatic display.
- ★ **Morning Glories (*Ipomoea spp.*):** These fast-growing vines produce colorful, trumpet-shaped flowers that open in the morning.

Ornamental Grasses:

- ★ **Pampas Grass (*Cortaderia* spp.):** Pampas grasses have tall, feathery plumes that add texture and movement to the landscape.
- ★ **Japanese Silver Grass (*Miscanthus* spp.):** These grasses have attractive foliage and feathery flower heads that sway in the breeze.

Annuals and Perennials:

- ★ **Geraniums (*Pelargonium* spp.):** Geraniums are popular bedding plants with a wide range of flower colors and patterns.
- ★ **Daffodils (*Narcissus* spp.):** Daffodils are classic spring-blooming bulbs known for their trumpet-shaped flowers.
- ★ **Begonias (*Begonia* spp.):** Begonias offer colorful flowers and foliage, making them suitable for both outdoor and indoor settings.

2
Flower Garden Design & Layout

Ornamental plants play a vital role in enhancing the visual appeal of various environments, whether it's a formal garden, urban landscape, or interior space. Their diverse forms, colors, and textures allow for endless creative possibilities in landscaping and design

Formal Garden

A formal garden is characterized by its structured and symmetrical design. It follows a set layout, often incorporating geometric shapes, straight lines, and meticulously planned arrangements of plants and hardscape elements. Here are some key features of a formal garden:

- **Symmetry:** Formal gardens prioritize symmetry, with plantings and pathways often mirrored on both sides of a central axis.
- **Clean Lines:** Straight paths, neatly trimmed hedges, and precisely aligned borders are common in formal garden designs.
- **Geometric Shapes:** Circular, square, or rectangular beds, often outlined with hedges or low walls, create a sense of order and precision.
- **Focal Points:** Formal gardens often have central focal points, such as statues, fountains, or elaborate flower beds.
- **Manicured Plants:** Plants are usually pruned and shaped to maintain a well-groomed appearance. Topiaries and closely clipped shrubs are common.
- **Color Coordination:** Colors are often carefully coordinated to create a harmonious and controlled color palette.

- ✯ **Hardscape Elements:** Paved pathways, symmetrical pergolas, and formal seating areas contribute to the structured look.

Informal Garden

An informal garden, also known as a cottage garden or naturalistic garden, embraces a more relaxed and organic style. It reflects a sense of spontaneity and often appears as if it developed naturally over time. Here are some key features of an informal garden:

- ✯ **Naturalistic Arrangement:** Informal gardens are designed to mimic the patterns found in nature, with a less rigid layout and asymmetrical features.
- ✯ **Mixed Plantings**: A variety of plants, including flowers, shrubs, and even native wildflowers, are combined in a way that appears less planned and more organic.
- ✯ **Curved Paths**: Paths and walkways may have a meandering, curving design that invites exploration and discovery.
- ✯ **Cottage-Style Plantings:** A mix of perennials, annuals, and self-seeding plants often contribute to the abundance of color and texture.
- ✯ **Relaxed Maintenance:** Informal gardens require less precise maintenance, allowing plants to grow more naturally and adapt to their environment.
- ✯ **Natural Focal Points:** Instead of formal statues or fountains, focal points may include elements like a bird bath, a seating area under a tree, or a rustic bench.
- ✯ **Color Harmony:** Colors are often allowed to mix more freely, giving the garden a vibrant and less controlled feel.
- ✯ **Wildlife Attraction:** Informal gardens often attract pollinators and wildlife due to the diverse plantings and relaxed environment.

Formal Flower Garden Layout

A formal garden layout is characterized by its structured and symmetrical design, often with geometric shapes and clean lines. Here's how to plan the layout:

Symmetry and Geometry:

- ★ Establish a central axis or focal point in the garden.
- ★ Use geometric shapes like squares, rectangles, and circles for pathways, flower beds, and hedges.

Pathways:

- ★ Create wide, straight pathways that intersect at right angles.
- ★ Use pavers, gravel, or formal materials for the paths.

Focal Points:

- ★ Place a central focal point, such as a decorative fountain, statue, or topiary, at the center of the garden.
- ★ Use symmetrically placed elements like ornamental containers, benches, or obelisks.

Flower Beds:

- ★ Design formal, raised flower beds with straight edges.
- ★ Plant symmetrically arranged flowers in patterns or blocks of color.

Hedges and Borders:

- ★ Use neatly trimmed hedges or low walls to define the edges of the garden and pathways.
- ★ Maintain uniform shapes for hedges, such as squares or rectangles.

Plant Selection:

- ★ Choose formal plants like roses, boxwood, and clipped evergreens.
- ★ Arrange plants in a structured manner, maintaining uniformity in spacing and heights.

Color Coordination:

- ★ Plan a coordinated color scheme with a limited palette for a polished look.

Informal Flower Garden Layout:

An informal garden layout embraces a more natural and relaxed style. Here's how to plan the layout for an informal garden:

Curved Pathways:

- ★ Create winding, meandering pathways that encourage exploration.
- ★ Use natural materials like gravel, stepping stones, or mulch.

Natural Arrangements:

- ★ Avoid strict symmetry and instead opt for asymmetrical arrangements of plants and features.

Mixed Plantings:

- ★ Combine a diverse selection of plants, including perennials, annuals, and native wildflowers.
- ★ Group plants based on their water and light requirements.

Focal Points:

- ★ Incorporate natural focal points like a seating area under a shade tree, a bird bath, or a whimsical sculpture.

Flower Beds and Borders:

- ★ Create free-form flower beds with organic, flowing edges.
- ★ Allow plants to spill over pathways and create a more relaxed boundary.

Planting Patterns:

- ★ Plant in a more naturalistic manner, allowing plants to intermingle and grow freely.

Color Variety:

- ★ Embrace a variety of colors and textures for a vibrant and dynamic garden.

Remember that your garden's layout should reflect your personal style, the site's characteristics, and your overall vision for the space. Whether formal or informal, the goal is to create a garden that brings joy and beauty to your outdoor environment.

Landscape ornamentals and shrubs play a crucial role in enhancing the beauty and functionality of outdoor spaces. They provide structure, texture, color, and often serve as focal points in gardens and landscapes.

3

Annuals and Perennial Flowers

In the vibrant tapestry of ornamental gardening, flowering annuals stand as the stars, painting the landscape with their brilliant hues and captivating forms. These transient treasures, celebrated for their ability to complete their life cycle in a single growing season, offer a captivating array of colors, shapes, and textures. In this chapter, we embark on a journey through the enchanting world of flowering annuals, uncovering their characteristics, cultivation techniques, and some of the most beloved varieties that adorn ornamental gardens.

The Allure of Flowering Annuals

- ✯ **Diversity in Color and Form:** Flowering annuals parade a kaleidoscope of colors, from delicate pastels to vibrant jewel tones. Their flowers vary in shape, size, and structure, offering an endless array of possibilities for creative garden design.
- ✯ **Ephemeral Beauty:** The fleeting nature of annuals' blooms adds a sense of preciousness to their splendor. Gardeners anticipate and savor their blossoms, knowing that their brilliance is fleeting but unforgettable.
- ✯ **Adaptability and Versatility:** Flowering annuals span various growth habits, including trailing, upright, and mounding forms, enabling them to thrive in beds, borders, containers, hanging baskets, and even as ground covers.

Characteristics of Flowering Annuals

Flowering annuals encompass a wide range of plant species with diverse characteristics:

- **Lifecycle**: Flowering annuals complete their life cycle in one growing season. They germinate from seeds, develop vegetatively, produce flowers, set seeds, and die, all within a single year.
- **Variety**: There is an astounding array of flowering annuals available, each with distinct colors, sizes, and shapes of flowers and foliage. This diversity enables gardeners to create dynamic displays.
- **Color and Aesthetics**: Flowering annuals are prized for their vibrant colors, which can range from delicate pastels to bold and bright hues. They can be used to create eye-catching focal points or harmonious color schemes.
- **Growth Habit**: Annuals can be classified into different growth habits such as trailing, mounding, upright, and climbing. This diversity allows gardeners to experiment with plant placement and arrangement.

The Cultivation of Floral Radiance

- **Site Selection and Soil Preparation:** Understanding the sunlight requirements of different annuals is crucial for optimal growth. Loamy, well-draining soil enriched with organic matter provides a nurturing medium for their roots to flourish.
- **Seeding and Transplanting:** Whether starting from seed indoors or directly sowing outdoors, proper timing is essential. Some annuals, like impatiens, require a head start indoors, while others, such as marigolds, can be directly sown in the garden.
- **Watering and Fertilization:** Consistent moisture is key to the success of annuals, especially during their establishment phase. Regular watering and controlled-release fertilizers promote healthy growth and abundant blooms.
- **Deadheading and Pruning:** To prolong the blooming period, deadheading—removing spent flowers—is recommended for many annuals. Additionally, some plants, like geraniums, benefit from occasional pinching to encourage bushier growth.
- **Pest and Disease Management:** Vigilant monitoring for pests and diseases allows for early intervention. Cultural practices such as proper spacing, good air circulation, and avoiding overhead watering can prevent many issues.

Exquisite Blooms: Notable Flowering Annuals

- **Petunia (*Petunia spp.*)**: Petunias are known for their trumpet-shaped flowers that come in a wide range of colors, including vibrant purples, pinks, blues, and whites. They can have solid, striped, or veined patterns on their petals. These compact plants are often used as bedding plants in flower beds, hanging baskets, and containers. They bloom profusely throughout the growing season and are loved for their ability to add a burst of color to any garden.
- **Marigold (*Tagetes spp.*)**: Marigolds are characterized by their aromatic foliage and bold, daisy-like flowers that typically come in shades of yellow and orange. They vary in size from low-growing to taller varieties. Marigolds are valued not only for their cheerful appearance but also for their ability to repel certain garden pests due to their strong scent. They are often used in flower beds, borders, and companion planting arrangements.
- **Zinnia (*Zinnia spp.*)**: Zinnias are versatile annuals with a wide range of flower forms and colors. They can have single or double petals and come in shades of red, pink, orange, yellow, and white. Zinnias are popular for their long-lasting blooms, making them excellent cut flowers. They attract butterflies and other pollinators, adding an interactive element to the garden. Zinnias work well in beds, borders, and cottage-style gardens.
- **Cosmos (*Cosmos spp.*)**: Cosmos plants feature delicate, daisy-like flowers with slender petals that come in colors such as pink, white, and various shades of red. Their airy and feathery foliage adds a touch of elegance to gardens. Cosmos are easy to grow from seed and thrive in sunny locations. They make great cut flowers and provide nectar for pollinators like bees and butterflies.
- **Alyssum (*Lobularia maritima*)**: Alyssum is a low-growing annual with clusters of tiny, fragrant flowers that come in shades of white, pink, and lavender. Their delicate appearance and sweet scent make them ideal for edging flower beds, borders, and rock gardens. Alyssum is also often used in container gardens and hanging baskets. They are known for attracting beneficial insects like hoverflies.
- **Snapdragon (*Antirrhinum majus*)**: Snapdragons are known for their distinctive, snap-like flowers that come in a variety of colors, including pastels and bold shades. The flowers are arranged in vertical spikes,

adding vertical interest to garden beds. Snapdragons are favored for their ability to bloom in cooler temperatures, making them suitable for both spring and fall gardens. They are often used as bedding plants and in mixed borders.

Flowering annuals are the painters of the garden canvas, infusing it with temporary but unforgettable beauty. Their versatility, dazzling array of colors, and varying forms offer a plethora of possibilities for gardeners seeking to create stunning, ever-changing displays. By mastering their cultivation techniques and understanding their unique characteristics, gardeners can weave these ephemeral gems into their ornamental landscapes, creating a living symphony of color and form.

Cultivation and Maintenance

Cultivating flowering annuals involves several key considerations:

- **Soil Preparation:** Well-draining, nutrient-rich soil is crucial for annuals. Amending soil with compost or organic matter improves drainage and fertility.
- **Sunlight Requirements:** Most flowering annuals thrive in full sun, although some tolerate partial shade. Properly assessing sunlight availability is essential for successful growth.
- **Watering:** Regular watering is important, especially during dry periods. However, overwatering can lead to root rot, so maintaining a balance is
  - **Petunia (*Petunia spp.*)**: Description: Petunias are known for their trumpet-shaped flowers that come in a wide range of colors, including vibrant purples, pinks, blues, and whites. They can have solid, striped, or veined patterns on their petals. These compact plants are often used as bedding plants in flower beds, hanging baskets, and containers. They bloom profusely throughout the growing season and are loved for their ability to add a burst of color to any garden.
  - **Marigold (*Tagetes spp.*)**: Description: Marigolds are characterized by their aromatic foliage and bold, daisy-like flowers that typically come in shades of yellow and orange. They vary in size from low-growing to taller varieties. Marigolds are valued not only for their cheerful appearance but also for their ability to repel certain garden pests due to their strong scent. They are often used in flower beds, borders, and companion planting arrangements.

- **Zinnia (*Zinnia spp.*)**: Description: Zinnias are versatile annuals with a wide range of flower forms and colors. They can have single or double petals and come in shades of red, pink, orange, yellow, and white. Zinnias are popular for their long-lasting blooms, making them excellent cut flowers. They attract butterflies and other pollinators, adding an interactive element to the garden. Zinnias work well in beds, borders, and cottage-style gardens.
- **Cosmos (*Cosmos spp.*)**: Description: Cosmos plants feature delicate, daisy-like flowers with slender petals that come in colors such as pink, white, and various shades of red. Their airy and feathery foliage adds a touch of elegance to gardens. Cosmos are easy to grow from seed and thrive in sunny locations. They make great cut flowers and provide nectar for pollinators like bees and butterflies.
- **Alyssum (*Lobularia maritima*)**: Description: Alyssum is a low-growing annual with clusters of tiny, fragrant flowers that come in shades of white, pink, and lavender. Their delicate appearance and sweet scent make them ideal for edging flower beds, borders, and rock gardens. Alyssum is also often used in container gardens and hanging baskets. They are known for attracting beneficial insects like hoverflies.
- **Snapdragon (*Antirrhinum majus*)**: Description: Snapdragons are known for their distinctive, snap-like flowers that come in a variety of colors, including pastels and bold shades. The flowers are arranged in vertical spikes, adding vertical interest to garden beds. Snapdragons are favored for their ability to bloom in cooler temperatures, making them suitable for both spring and fall gardens. They are often used as bedding plants and in mixed borders.

Herbaceous Perennials in Ornamental Landscaping

Herbaceous perennials are the backbone of many ornamental landscapes, providing structure, texture, and consistent beauty year after year. These plants, characterized by their non-woody stems and longevity, offer a diverse palette of colors, forms, and growth habits. In this chapter, we will delve into the world of herbaceous perennials, exploring their characteristics, cultivation, and noteworthy species for creating stunning and enduring garden displays.

Characteristics of Herbaceous Perennials

Herbaceous perennials encompass a wide array of plants with distinct attributes:

- **Longevity**: Unlike annuals, herbaceous perennials live for multiple years, often regrowing from the same root system each spring.
- **Variety:** From low-growing ground covers to tall and dramatic specimens, herbaceous perennials come in various sizes, shapes, and growth habits, allowing for creative landscaping designs.
- **Seasonal Interest:** Many perennials offer multi-seasonal interest, with varying bloom times, foliage colors, and seed heads, providing beauty throughout the year.
- **Adaptability:** Herbaceous perennials exhibit a broad range of adaptability to different climates, soils, and light conditions, making them suitable for various garden settings.

Cultivation and Maintenance

Growing herbaceous perennials successfully involves the following key considerations:

- **Site Selection:** Assessing sunlight, soil type, drainage, and microclimates within the garden helps determine the best placement for each perennial species.
- **Soil Preparation:** Improving soil structure and fertility through practices like adding organic matter and proper pH adjustment supports healthy growth.
- **Planting Techniques:** Planting perennials at the correct depth and spacing, and incorporating proper planting practices, such as watering in well, encourages strong root establishment.
- **Mulching:** Applying a layer of organic mulch helps retain soil moisture, suppress weeds, and maintain more consistent soil temperatures.
- **Watering:** Adequate watering during establishment and proper irrigation thereafter is crucial, with attention to each plant's moisture preferences.
- **Division:** Many perennials benefit from periodic division to rejuvenate and prevent overcrowding. Dividing plants every few years maintains vigor and bloom quality.

- **Pruning:** Regular dead heading and, when appropriate, cutting back after flowering or in late fall helps shape plants, prevent self-seeding, and promotes better growth.

Herbaceous Perennials

- **Hosta (*Hosta spp.*)**: Hostas are known for their striking foliage, which comes in a variety of shapes, sizes, and colors, including shades of green, blue, yellow, and variegated combinations. They thrive in shaded to partially shaded areas and are often used for creating lush ground covers or adding texture to shady borders. Some hostas also produce tall spikes of bell-shaped flowers in summer.
- **Daylily (*Hemerocallis spp.*)**: Daylilies are cherished for their vibrant and varied blooms, which appear in a wide range of colors, including reds, oranges, yellows, and purples. Each flower typically lasts only a day, but multiple blooms open over the course of several weeks, creating a continuous display. They are versatile perennials that work well in sunny to partially shaded locations and are often used as border plants or in mixed beds.
- **Black-eyed Susan (*Rudbeckia spp.*)**: Black-eyed Susans are iconic with their golden-yellow, daisy-like flowers that feature a dark center, resembling a "black eye." They bring a touch of warmth to gardens and attract pollinators like bees and butterflies. These sun-loving perennials are commonly used in wildflower gardens, meadows, and cottage-style landscapes.
- **Peony (*Paeonia spp.*)**: Peonies are known for their luxurious, large, and fragrant blooms that come in a variety of colors, including white, pink, red, and even coral. They have a classic and timeless appeal and are often associated with traditional gardens. Peonies prefer full sun to light shade and are treasured for their ability to produce stunning floral displays year after year.
- **Sedum (*Sedum spp.*)**: Sedums, also known as stonecrop, are drought-tolerant perennials valued for their fleshy leaves and star-like flowers. They come in a range of sizes, from low-growing ground covers to tall upright varieties. Sedums often feature succulent foliage and offer varying flower colors, attracting pollinators like bees and butterflies. They are ideal for rock gardens, sunny borders, and containers.
- **Lavender (*Lavandula spp.*)**: Lavender is prized for its aromatic gray-green foliage and spikes of small, fragrant flowers that range in color

from pale purple to deep blue. Lavender thrives in full sun and well-drained soil, making it a popular choice for herb gardens, borders, and as a fragrant addition to landscape designs. It also has medicinal and culinary uses.

These examples represent just a glimpse into the diverse world of herbaceous perennials. Each species offers unique characteristics that contribute to the beauty and charm of ornamental gardens, creating visually captivating and enduring landscapes.

Herbaceous perennials are the cornerstone of long-lasting, dynamic ornamental landscapes. Their enduring presence, combined with their diversity of forms and colors, offers gardeners the opportunity to create captivating displays that evolve with the seasons. By understanding their cultivation needs and maintenance requirements, gardeners can harness the potential of herbaceous perennials to craft beautiful, resilient gardens that stand the test of time.

4

Ornamental Bulbous Plants: A Symphony of Color and Form

Ornamental bulbous plants have been captivating garden enthusiasts and horticulturists for centuries with their exquisite beauty and remarkable adaptability. These unique plants store their energy and nutrients in underground bulb structures, allowing them to survive adverse conditions and burst forth in a riot of colors and shapes when the time is right.

Characteristics of Ornamental Bulbous Plants

Ornamental bulbous plants belong to various botanical families and encompass a wide range of species, each with its own distinctive characteristics. Common features include:

- **Bulb Structure:** The bulb is the heart of these plants, containing the embryonic shoot, modified leaves, and stored nutrients. This structure allows them to survive unfavorable conditions, including drought and cold temperatures.
- **Flower Diversity:** Ornamental bulbous plants exhibit an astonishing array of flower forms, sizes, and colors. From delicate, nodding snowdrops (*Galanthus spp.*) to the flamboyant trumpet-shaped blooms of daffodils (*Narcissus spp.*), the diversity is unparalleled.
- **Seasonal Beauty:** These plants are known for their seasonal displays, often heralding the arrival of spring with their vibrant blossoms. However, some species, such as autumn crocuses (*Crocus spp.*), bring their charm to the fall garden.

- **Ease of Cultivation:** Many ornamental bulbous plants are relatively easy to grow, making them ideal for both novice and experienced gardeners. Their adaptability to various soil types and climates contributes to their popularity.

Cultivation and Care

Successful cultivation of ornamental bulbous plants requires attention to several key factors:

- **Site Selection**: Choose a location with well-draining soil to prevent bulb rot. Most bulbous plants thrive in areas with ample sunlight, although some, like woodland natives, prefer dappled shade.
- **Planting Depth**: Proper planting depth is critical. Generally, bulbs should be planted at a depth equal to two to three times their diameter. Deeper planting is recommended for larger bulbs to ensure stability.
- **Watering:** Bulbs require consistent moisture during their growth phase. However, excessive waterlogging can lead to root rot. Adjust watering practices based on the specific needs of the plant.
- **Fertilization:** Prior to planting, amend the soil with a balanced, slow-release fertilizer. Avoid excessive nitrogen, which can lead to lush foliage but fewer flowers.
- **Naturalization:** Some bulbous plants, like crocuses and snowdrops, are well-suited for naturalizing in lawns or woodland areas. Allow foliage to die back naturally to replenish bulb nutrients.

Ornamental Bulbous Plants

- **Tulips (*Tulipa spp.*)**: With over 3,000 registered varieties, tulips are a symbol of spring. Their cup-shaped flowers come in an array of colors, heights, and bloom times, providing endless design possibilities.
- **Hyacinths (*Hyacinthus orientalis*)**: These fragrant blooms form dense spikes of flowers in captivating hues. Hyacinths are perfect for containers and garden beds alike.
- **Alliums (*Allium spp.*)**: Ranging from small, drumstick-like blooms to towering spheres, alliums add architectural interest to gardens. Their distinctive forms and shades make them standout choices.
- **Crocuses (*Crocus spp.*)**: Often the first to bloom in spring, crocuses carpet the ground with jewel-toned flowers. They're also available in fall-blooming varieties.

- ★ **Daffodils (*Narcissus spp.*)**: Daffodils come in a spectrum of colors and shapes, from classic yellow trumpets to intricate multi-petaled varieties. They naturalize easily.
- ★ **Lilies (*Lilium spp.*)**: Though not true bulbs, lilies are often grouped with bulbous plants. Their stunning, large flowers and diverse types, such as Asiatic and Oriental lilies, are garden showstoppers.

Ornamental bulbous plants hold an esteemed place in gardening history, captivating generations with their ephemeral yet awe-inspiring beauty. From their distinctive bulb structures to the vast array of flower forms, they provide endless opportunities for creativity in garden design. By understanding their characteristics and needs, both beginners and seasoned gardeners can cultivate these botanical treasures and enjoy their remarkable displays for years to come.

5

Elegance in Resilience: Ornamental Cacti & Succulents

In the arid landscapes around the world, a fascinating array of plants thrives, captivating the hearts of plant enthusiasts and designers alike. Ornamental cacti and succulents, masterpieces of adaptation, have found ingenious ways to store water and withstand harsh conditions. This chapter delves into the enchanting realm of these resilient plants, exploring their unique features, cultivation secrets, and showcasing remarkable examples that grace both indoor and outdoor spaces.

A World of Diversity: Cacti and Succulents

Ornamental cacti and succulents are living sculptures, characterized by their water-storing abilities, architectural forms, and captivating textures. They've gained immense popularity for their:

- **Water-Storing Adaptations:** Both cacti and succulents have evolved to retain water in their fleshy tissues, allowing them to survive in arid environments where water is scarce.
- **Diverse Forms:** From the tall, columnar shapes of saguaro cacti (*Carnegiea gigantea*) to the compact rosettes of echeverias (*Echeveria spp.*), these plants display an astonishing variety of shapes, sizes, and growth habits.
- **Leaf Texture and Color:** Succulents are celebrated for their fascinating leaf textures, ranging from smooth and glossy to spiky and waxy. The color spectrum they encompass—shades of green, blue, purple, and even red—adds to their visual allure.

- ✯ **Ease of Care:** Cacti and succulents are well-suited for both novice and experienced gardeners due to their minimal water requirements and ability to thrive in various light conditions.

Cultivating and Caring for Cacti and Succulents

To cultivate healthy and vibrant cacti and succulents, gardeners should consider the following factors:

- ✯ **Light Requirements:** While most cacti require bright sunlight, some succulents thrive in lower light conditions. Carefully observe light needs to prevent sunburn or etiolation.
- ✯ **Potting Mix and Containers:** Well-draining soil is crucial for preventing root rot. Use a specialized cactus or succulent mix, and ensure containers have drainage holes.
- ✯ **Watering Practices:** Allow the soil to dry out between waterings, as overwatering is a common issue. Adjust watering frequency based on the plant's specific needs and environmental conditions.
- ✯ **Temperature Tolerance:** While cacti are often associated with desert environments, many succulents originate from a range of climates. Be aware of temperature requirements for each plant.
- ✯ **Propagation:** Cacti and succulents can be propagated from seeds, offsets, cuttings, or even leaves. Understanding the propagation methods of each species is essential for successful reproduction.

Remarkable Examples of Ornamental Cacti and Succulents

- ✯ **Aloe Vera (*Aloe barbadensis miller*)**: Renowned for its healing properties, aloe vera's succulent leaves contain soothing gel. Its architectural rosette form adds a touch of elegance.
- ✯ **Golden Barrel Cactus (*Echinocactus grusonii*)**: This globular cactus features a distinctive appearance, with numerous golden spines radiating from its center.
- ✯ **Zebra Plant (*Haworthiopsis attenuata*)**: Characterized by its striking zebra-like patterns on its leaves, this succulent is popular for its low-maintenance care.

- **Pincushion Cactus (*Mammillaria spp.*)**: These small cacti are known for their intricate clusters of spines and colorful flowers. Their compact size makes them suitable for small spaces.
- **String of Pearls (*Senecio rowleyanus*)**: A trailing succulent with bead-like leaves, this plant adds a unique cascading element to hanging baskets or shelves.
- **Agave (*Agave spp.*)**: Ranging from large architectural species to smaller, more compact varieties, agaves are iconic succulents known for their dramatic rosettes and spiky leaves.

Ornamental cacti and succulents embody the elegance of survival in challenging environments. Their unique forms, textures, and adaptability have made them sought-after additions to landscapes and interiors alike. By understanding their specific needs and showcasing their beauty, gardeners can enjoy the beauty of these fascinating plants while celebrating their resilience.

6

Cultivation of Cut Flowers

Aster (Asteraceae)

Asters, belonging to the Asteraceae family, are popular and versatile cut flowers known for their daisy-like appearance and vibrant colors. These flowers are beloved for their long-lasting blooms and ability to add charm to various floral arrangements. This guide provides an in-depth look at the cultivation of asters, offering insights and tips to help you grow healthy plants and achieve a bountiful harvest for stunning cut flower displays.

Varieties of Asters for Cut Flowers: There are numerous aster varieties available for cut flower cultivation, each with its unique characteristics. Some popular aster varieties for cut flowers include:

- Aster novi-belgii (New York Aster)
- Aster novae-angliae (New England Aster)
- Aster amellus (Italian Aster)
- Aster laevis (Smooth Aster)
- Aster ericoides (Heath Aster)

Cultivation Guidelines

- **Site Selection:** Asters thrive in full sunlight, so choose a well-drained location that receives at least 6 to 8 hours of direct sunlight daily.

- ★ **Soil Preparation:** Prepare the soil by incorporating organic matter to improve drainage and fertility. A slightly acidic to neutral pH (6.0-7.0) is ideal for aster cultivation.
- ★ **Planting:** Plant aster seeds or young plants after the last frost date in your area. Space plants according to the variety's specific requirements, usually around 12 to 18 inches apart.
- ★ **Watering:** Asters prefer consistent moisture, but they don't tolerate waterlogged soil. Water at the base of the plants to avoid wetting the foliage, which can lead to disease.
- ★ **Fertilization:** Apply a balanced, slow-release fertilizer before planting and again during the growing season to encourage healthy growth and abundant blooms.
- ★ **Support:** Some aster varieties may benefit from staking to prevent bending or breaking of stems under the weight of flowers. Place supports early in the season to avoid damaging established roots.
- ★ **Pruning:** Pinch back young plants when they are about 6 inches tall to encourage branching and compact growth. Deadhead spent flowers to prolong blooming.
- ★ **Pest and Disease Management:** Keep an eye out for aphids, caterpillars, and powdery mildew. Utilize natural predators or appropriate organic treatments to manage pests and diseases.
- ★ **Overwintering:** Apply a layer of mulch around the base of the plants in late fall to protect the roots from extreme cold. Asters are hardy perennials in many regions.

Harvesting Asters:

Asters are best harvested when about 50-75% of the flowers on a stem are open. Cut stems in the early morning or late afternoon to minimize stress on the plants. Place the cut stems in a bucket of water immediately to prevent wilting.

By following these cultivation guidelines and leveraging the references provided, you can embark on a successful journey to grow stunning asters that will grace your floral arrangements with their vibrant colors and enduring beauty.

Cultivation of Cut Flowers: Dahlias (Dahlia spp.)

Dahlias are renowned for their diverse forms, vibrant colors, and impressive size range, making them a favorite among cut flower enthusiasts. With proper cultivation, dahlias can produce an abundance of blooms that add elegance and charm to floral arrangements. This guide offers valuable insights into the cultivation of dahlias for cut flowers, providing tips and references to help you achieve spectacular results.

Varieties of Dahlias for Cut Flowers: Dahlias come in various sizes, shapes, and colors, allowing for a rich assortment in cut flower arrangements. Some popular dahlia types for cut flowers include:

- **Decorative Dahlias:** Large, fully double flowers with flat petals.
- **Cactus Dahlias:** Rolled or spiky petals that create a striking texture.
- **Pompon Dahlias:** Small, ball-shaped flowers with tightly packed petals.
- **Anemone Dahlias:** A central disc surrounded by a single or double row of petals.
- **Waterlily Dahlias:** Petals arranged in a loose, waterlily-like fashion.

Cultivation Guidelines

- **Site Selection:** Choose a sunny location with well-draining soil. Dahlias thrive in full sun, receiving at least 6 to 8 hours of direct sunlight daily.
- **Soil Preparation:** Work organic matter into the soil to improve its structure and fertility. Dahlias prefer a slightly acidic to neutral pH (6.5-7.0).
- **Planting:** Plant dahlia tubers after the danger of frost has passed. Space tubers according to the variety's size, usually 18 to 36 inches apart.
- **Watering:** Keep the soil consistently moist but not waterlogged. Water at the base of the plants to prevent fungal diseases.
- **Fertilization:** Incorporate a balanced, slow-release fertilizer into the soil before planting. Supplement with a high-phosphorus fertilizer during the growing season to promote flowering.
- **Support:** As dahlias grow tall and produce heavy blooms, staking is essential to prevent bending or breaking. Insert stakes at planting or when plants are small to avoid damaging roots.
- **Pruning:** Pinch back the terminal shoot when plants have 3 to 4 pairs of leaves to encourage branching. Deadhead spent blooms to encourage continuous flowering.

- ✯ **Pest and Disease Management:** Monitor for aphids, slugs, and powdery mildew. Introduce natural predators or use appropriate treatments to manage pests and diseases.
- ✯ **Overwintering:** In regions with frost, dig up tubers after the first frost kills the foliage. Store tubers in a cool, dry location until the next planting season.

Harvesting Dahlias:

Dahlias are best harvested when the blooms are fully open but not overmature. Use sharp pruners to cut stems at an angle, and place them immediately in a bucket of water to prevent wilting.

Cultivation of Cut Flowers: Gerbera (*Gerbera jamesonii*)

Introduction: Gerberas, with their bold and vibrant petals, are highly sought-after cut flowers that bring a touch of elegance to any floral arrangement. Their wide range of colors and shapes makes them a versatile choice for various occasions. This guide offers valuable insights into the cultivation of gerberas for cut flowers, providing tips and references to help you achieve stunning results.

Varieties of Gerberas for Cut Flowers: Gerberas are available in numerous colors and sizes, making them a favorite choice for florists and flower enthusiasts alike. Some popular gerbera varieties for cut flowers include:

- ✯ **Single Flowered Gerberas:** Large, single blooms with a prominent central disc.
- ✯ **Double Flowered Gerberas:** Flowers with multiple rows of petals, creating a lush appearance.
- ✯ **Miniature Gerberas:** Compact plants with smaller blooms, perfect for petite arrangements.
- ✯ **Crested Gerberas:** Flowers with unique and intricate formations in the central disc.

Cultivation Guidelines:

- ✯ **Site Selection:** Choose a site with well-draining soil and good air circulation. Gerberas prefer bright, indirect light or filtered sunlight.
- ✯ **Soil Preparation:** Ensure the soil is well-draining and rich in organic matter. A slightly acidic to neutral pH (6.0-7.0) is ideal for gerbera cultivation.
- ✯ **Planting:** Plant gerbera plants or seeds after the last frost date. Space plants according to their mature size, usually 12 to 18 inches apart.

- **Watering:** Keep the soil consistently moist but not waterlogged. Water at the base of the plant to prevent fungal diseases.
- **Fertilization:** Apply a balanced, slow-release fertilizer during the growing season to promote healthy growth and flowering.
- **Temperature and Humidity:** Gerberas thrive in moderate temperatures and appreciate humidity. Avoid exposing them to extreme heat or cold.
- **Pest and Disease Management:** Monitor for aphids, spider mites, and fungal issues. Utilize natural predators or appropriate treatments to manage pests and diseases.
- **Deadheading:** Regularly remove spent blooms to encourage continuous flowering and maintain plant aesthetics.
- **Overwintering:** In areas with frost, gerberas are often grown as annuals or potted plants. In warmer regions, they can be grown as perennials.

Harvesting Gerberas:

Gerberas are best harvested when the outer petals have fully opened and the center disc is visible. Use sharp scissors or pruners to cut stems at an angle, and place them immediately in a container of water to prevent wilting.

Anthuriums (*Anthurium* spp.)

Anthuriums, often referred to as "flamingo flowers" or "laceleaf," are prized for their unique, heart-shaped blooms and glossy foliage. These tropical beauties have become popular choices for cut flower arrangements due to their striking appearance and long vase life. This guide provides valuable insights into the cultivation of anthuriums for cut flowers, offering tips and references to help you cultivate these elegant blooms successfully.

Varieties of Anthuriums for Cut Flowers: Anthuriums come in various colors and shapes, offering a diverse array of options for cut flower enthusiasts. Some popular anthurium varieties for cut flowers include:

- ***Anthurium andraeanum:*** Known for its waxy, heart-shaped spathes in shades of red, pink, or white.
- ***Anthurium scherzerianum***: Recognized for its bright red, spikelike spathes and contrasting yellow spadices.
- ***Anthurium hybrid***: A range of hybrids that exhibit a variety of spathes and spadix colors, including shades of purple, green, and bicolor combinations.

Cultivation Guidelines:

- ★ **Site Selection:** Choose a location with bright, indirect light or filtered sunlight. Avoid direct sunlight, as it can scorch the leaves.
- ★ **Soil Preparation:** Use a well-draining, porous potting mix that retains moisture without becoming waterlogged.
- ★ **Planting:** Anthuriums are typically grown from nursery plants. Ensure proper spacing to allow air circulation and prevent overcrowding.
- ★ **Watering:** Keep the soil consistently moist but not soggy. Allow the top inch of the soil to dry between waterings.
- ★ **Humidity and Temperature:** Anthuriums thrive in high humidity and warm temperatures. Maintain a humidity level of around 60-80%.
- ★ **Fertilization:** Apply a balanced liquid fertilizer every 6-8 weeks during the growing season (spring and summer). Avoid over-fertilization.
- ★ **Pruning:** Remove any yellowing or damaged leaves to maintain the plant's appearance and health.
- ★ **Pest and Disease Management:** Watch for pests like aphids, mealybugs, and spider mites. Treat infestations promptly using appropriate measures.
- ★ **Potting:** Repot anthuriums when they outgrow their containers, usually every 2-3 years. Use a slightly larger pot and fresh potting mix.

Harvesting Anthuriums:

Anthuriums are best harvested when the spathes are fully open and the spadices are erect. Use sharp scissors or pruners to cut stems, making clean cuts to prevent damage to the plant. Place the cut stems immediately in water to maintain their freshness.

Marigold

Marigold (*Tagetes spp.)* is a popular cut flower that is widely cultivated for its vibrant colors, long vase life, and pleasant fragrance. It is commonly used in floral arrangements, garlands, and decorations. Marigold cultivation is relatively easy and can be adapted to various climates. Here's a basic guide to cultivating marigold cut flowers:

Varieties of Marigolds for Cut Flowers: Marigolds come in various sizes and colors, offering a wide range of options for cut flower enthusiasts. Some popular marigold varieties for cut flowers include:

- ★ **African Marigolds (*Tagetes erecta*):** Taller plants with large, double blooms in shades of yellow, orange, and red.
- ★ **French Marigolds (*Tagetes patula*):** Compact plants with small, delicate blooms in vibrant hues.
- ★ **Signet Marigolds (*Tagetes tenuifolia*):** Edible flowers with small, single blooms and finely divided foliage.

Cultivation of Marigold (*Tagetes spp.*) as cut flowers can be a rewarding venture due to their vibrant colors, long vase life, and ease of cultivation. Marigolds are popular choices for floral arrangements and decorations. Below is a guide to cultivating Marigold as cut flowers, along with some references for further reading:

Varieties: Choose Marigold varieties that are well-suited for cut flower production. Some popular varieties include:

- ★ **African Marigolds (*Tagetes erecta*):** Taller plants with large, double flowers.
- ★ **French Marigolds (*Tagetes patula*):** Compact plants with smaller, double or single flowers.
- ★ **Site Selection:** Choose a sunny location with well-draining soil. Marigolds prefer full sun to thrive and produce abundant blooms.
- ★ **Soil Preparation:** Prepare the soil by adding organic matter such as compost to improve soil structure and fertility. Ensure good drainage to prevent waterlogged conditions.
- ★ **Sowing:** Marigolds can be directly sown into the garden or started indoors and later transplanted. Sow seeds in well-prepared soil after the last frost date. Follow the spacing recommendations on the seed packet, as spacing can vary depending on the variety.
- ★ **Transplanting:** If starting indoors, sow seeds in trays or pots several weeks before the last frost. Transplant seedlings outdoors once the danger of frost has passed and seedlings have developed a few sets of true leaves.

Care:

- ★ **Watering:** Keep the soil evenly moist but not waterlogged. Avoid overhead watering, as wet foliage can lead to fungal diseases.
- ★ **Fertilization:** Apply a balanced fertilizer according to the package instructions to promote healthy growth and flower production.

- ✯ **Mulching:** Apply a layer of organic mulch around the plants to conserve soil moisture and suppress weed growth.
- ✯ **Pinching:** Pinch off the terminal bud when plants are young to encourage branching and the development of more flower buds.
- ✯ **Deadheading:** Remove spent flowers regularly to encourage continuous blooming. This process, known as deadheading, redirects the plant's energy into producing new blooms.
- ✯ **Pest and Disease Control:** Monitor plants for common pests such as aphids, spider mites, and whiteflies. Implement appropriate pest control measures if needed. Also, be vigilant for diseases like powdery mildew and rust, and treat accordingly.
- ✯ **Harvesting:** Harvest marigold blooms when they are fully open but before they start to decline. Cut stems early in the morning or late in the afternoon to minimize water loss. Use sharp, clean scissors or pruning shears.
- ✯ **Vase Life:** Marigolds have a relatively good vase life. To extend their vase life, place them in clean water, remove any foliage that will be submerged in the water, and change the water regularly.

Lilium

Lilies are elegant and popular cut flowers known for their striking appearance and pleasant fragrance. Cultivating lilies for cut flower production requires careful attention to planting, care, and harvesting. Here's a comprehensive guide to cultivating Lilium for cut flowers:

- ✯ **Selection of Varieties:** Choose lily varieties that are suitable for cut flower production. Common cut flower lily types include Oriental lilies, Asiatic lilies, Trumpet lilies, and LA (Longiflorum-Asiatic) hybrids. Consider factors such as color, size, and bloom time.
- ✯ **Site Selection:** Select a planting site with well-draining soil and access to sunlight. Lilies generally prefer full sun to light shade. Ensure the site has good air circulation to prevent fungal diseases.
- ✯ **Soil Preparation:** Prepare the soil by adding organic matter such as compost to improve soil structure and fertility. Lilies prefer slightly acidic to neutral soil pH (around 6.0 to 7.0).

Planting:

- ★ **Time:** Lilies can be planted in the fall or spring. Fall planting is preferred as it allows the bulbs to establish roots before the growing season.
- ★ **Depth:** Plant bulbs at a depth of about 3 times the height of the bulb.
- ★ **Spacing:** Space bulbs according to the specific variety's recommendations, usually around 6 to 12 inches apart.

Care:

- ★ **Watering:** Keep the soil consistently moist, especially during active growth and flowering. Avoid overwatering, which can lead to bulb rot.
- ★ **Fertilization:** Apply a balanced fertilizer in the spring as new growth emerges. Avoid high-nitrogen fertilizers, as they can result in excessive foliage growth at the expense of flowers.
- ★ **Mulching:** Apply a layer of mulch to help retain soil moisture and suppress weed growth.
- ★ **Staking:** Tall lily varieties may require staking to support the stems and prevent them from bending or breaking under the weight of the flowers. Use stakes or other supports to gently secure the stems.
- ★ **Pest and Disease Control:** Monitor for pests such as aphids, lily beetles, and slugs. Employ appropriate pest control measures if needed. Watch for diseases like botrytis and rust and take preventive actions.
- ★ **Harvesting:** Harvest lily flowers when they are in the bud stage or have just started to open. The buds will continue to open after cutting. Use clean, sharp shears or knives to cut stems at a slight angle.
- ★ **Vase Life:** Lilies have a relatively good vase life. Remove any leaves that will be submerged in water to prevent bacterial growth. Change the vase water every few days to extend their vase life.
- ★ **Storage:** If you need to store lily blooms before arranging, place them in a cool, dark location with high humidity. This can help extend their freshness.

7

Cultivating Plants in Pots and the Art of Indoor Gardening

Indoor gardening offers a unique opportunity to transform living spaces into vibrant, thriving oases of greenery. Cultivating plants in pots brings the beauty of the natural world indoors, providing solace, aesthetic appeal, and even improved air quality. This chapter explores the art of indoor gardening, delving into the intricacies of pot cultivation and highlighting a selection of plants that flourish in indoor environments.

Embracing Indoor Gardening: The Power of Potted Plants

Indoor gardening goes beyond mere decoration; it fosters a connection to nature, even within urban environments. Cultivating plants in pots within the home offers a range of benefits:

- **Enhanced Aesthetics**: Potted plants introduce colors, shapes, and textures that enhance interior design, creating an inviting and visually appealing atmosphere.
- **Healthier Indoor Air**: Many indoor plants actively improve air quality by filtering out pollutants and releasing oxygen, contributing to a healthier living environment.
- **Wellness and Relaxation**: The presence of greenery has been shown to reduce stress and boost overall well-being, creating a calming oasis within the home.

- **Flexibility and Creativity**: Potted plants allow for creative expression and experimentation, enabling homeowners to craft personalized indoor gardens.

Cultivating Plants in Pots

Cultivating plants in pots requires a tailored approach to ensure optimal growth and health:

- **Choosing the Right Containers:** Select pots with drainage holes to prevent waterlogging. Choose sizes appropriate for the plant's growth and consider aesthetics that complement your indoor space.
- **Selecting the Ideal Soil:** Use well-draining potting mix tailored to the specific plant's needs. Add perlite or vermiculite for additional aeration.
- **Light and Placement:** Assess the light conditions of your indoor space to choose plants that match. Consider factors like south-facing windows for bright light or low-light corners for shade-loving plants.
- **Watering Wisdom:** Establish a consistent watering routine while allowing the top inch or so of soil to dry out between waterings. Adjust watering frequency based on the plant's water requirements and the indoor environment.
- **Humidity and Temperature:** Monitor indoor humidity and temperature levels, especially during seasons with extreme conditions. Some plants may benefit from occasional misting or grouping to increase humidity.

Remarkable Plants for Indoor Gardens

- **Snake Plant (*Sansevieria spp.*)**: This hardy plant thrives in low light and efficiently purifies the air by removing toxins.
- **Spider Plant (*Chlorophytum comosum*)**: Known for its arching, variegated leaves, the spider plant is easy to care for and produces baby plantlets that can be propagated.
- **Peace Lily (*Spathiphyllum spp.*)**: With its elegant white flowers and air-purifying abilities, the peace lily brings a touch of tranquility to indoor spaces.
- **ZZ Plant (*Zamioculcas zamiifolia*)**: This striking plant boasts glossy, dark green leaves and thrives in low-light conditions, making it a perfect choice for dimly lit rooms.

- **Pothos (*Epipremnum aureum*)**: An excellent trailing plant, pothos comes in various leaf colors and is known for its adaptability and air-purifying qualities.
- **Fiddle Leaf Fig (*Ficus lyrata*)**: With its large, glossy leaves, the fiddle leaf fig adds a touch of drama and sophistication to indoor spaces.

Cultivating plants in pots and practicing indoor gardening is a rewarding endeavor that infuses indoor spaces with the beauty and vitality of the natural world. By understanding the specific needs of potted plants and selecting the right species for your environment, you can create an indoor garden that not only enhances the aesthetics of your home but also contributes to your well-being.

Factors Influencing Flower Production

The beauty of flowers gracing gardens, fields, and landscapes is a mesmerizing sight that captures the essence of nature's creativity. However, the journey from bud to blossom is influenced by an intricate interplay of environmental, physiological, and genetic factors. This chapter delves into the fascinating world of flower production, examining the key elements that shape the development and abundance of blooms. Through understanding these factors, gardeners and enthusiasts can nurture flourishing floral displays that enchant the senses.

The Complex Orchestra of Flower Production

Flower production involves a symphony of factors, each playing its own distinct role in the creation and development of blooms. These factors include:

- **Light and Photoperiod:** The amount and duration of light exposure significantly impact flower initiation. Photoperiod-sensitive plants rely on specific day-length conditions to trigger the transition from vegetative growth to flowering.
- **Temperature and Climate:** Temperature influences flowering in various ways. Some plants require a period of chilling (vernalization) to induce flowering, while others thrive under warm conditions.
- **Nutrition and Soil Health:** Proper nutrient availability, especially phosphorus and potassium, is essential for flower production. Imbalances can lead to poor flowering or excessive vegetative growth.

- **Water and Moisture:** Adequate and consistent watering supports healthy flower development. Drought stress can disrupt flower formation, while overwatering might cause root rot and inhibit blooming.
- **Pruning and Deadheading:** Pruning practices can impact flowering by affecting the allocation of energy between growth and reproduction. Deadheading spent flowers can encourage continuous blooming.

Influences on Flowering: Exploring Examples

- **Lilacs (*Syringa spp.*)**: These deciduous shrubs require a period of winter chill to produce blooms in spring. Insufficient chilling can result in reduced flowering.
- **Poinsettias (*Euphorbia pulcherrima*)**: Poinsettias exhibit photoperiod sensitivity, requiring long nights to trigger colorful bract development. They're popular choices for holiday displays.
- **Daylilies (*Hemerocallis spp.*)**: Daylilies are versatile bloomers, but their flower production can be influenced by soil fertility and moisture levels.
- **Tomatoes (*Solanum lycopersicum*)**: While commonly grown for their fruit, tomatoes produce flowers that develop into fruits. Proper nutrition and environmental conditions contribute to flower production.
- **Chrysanthemums (*Chrysanthemum spp.*)**: Chrysanthemums require specific photoperiods for floral initiation, making them seasonal bloomers often associated with autumn.
- **Roses (*Rosa spp.*)**: Pruning practices play a vital role in rose flowering. Proper deadheading encourages more blooms, while severe pruning might delay flowering.

The process of flower production is a delicate balance of external and internal factors, all intricately linked to the plant's growth and reproductive mechanisms. Gardeners who recognize and respect these factors can orchestrate vibrant floral displays that harmonize with the rhythms of nature. By understanding the nuances of light, temperature, nutrition, and other variables, one can cultivate gardens that burst forth in a spectacular symphony of color and fragrance.

Crafting Beauty: Production and Packaging of Cut Flowers

Cut flowers, a symbol of love, celebration, and remembrance, hold the power to evoke emotions through their exquisite forms and fragrances. Behind the enchanting blooms that grace homes, events, and businesses lies a meticulous process of production and packaging. This chapter explores the art and science of producing and presenting cut flowers, shedding light on the techniques that ensure the longevity and beauty of these ephemeral treasures.

From Field to Vase: The Journey of Cut Flowers

Producing cut flowers is a collaborative effort between nature and human intervention. Key stages include:

- **Cultivation and Harvesting**: Flowers are grown in fields or greenhouses, carefully nurtured to ensure optimal growth and bloom. The timing of harvesting plays a crucial role in determining flower quality and vase life.
- **Post-Harvest Handling**: Flowers are harvested early in the morning when they are fully hydrated. After harvest, they undergo processing steps such as hydration, cleaning, and leaf removal.
- **Packaging and Transportation**: Flowers are carefully packaged to prevent damage during transit. Packaging materials include protective sleeves, wraps, and hydration solutions to maintain moisture levels.
- **Market Distribution**: Flowers are transported to wholesalers, retailers, and even directly to consumers. Careful handling and temperature control are essential to preserving freshness.

Prolonging Beauty: Techniques for Extending Vase Life

Flowers have a finite vase life, but various methods can extend their beauty:

- **Hydration**: Rehydrating cut flowers immediately upon arrival helps prevent wilting. Trim stems at an angle and place them in clean, warm water containing flower preservatives.
- **Temperature Control**: Flowers are sensitive to temperature fluctuations. Display them away from direct sunlight, drafts, and heat sources to prolong freshness.

- **Water Quality**: Using clean, fresh water is crucial. Change the water every few days and re-cut stems to prevent bacterial growth.
- **Flower Food**: Commercial flower preservatives provide essential nutrients and inhibit bacterial growth, extending vase life.

Exemplary Cut Flowers: A Showcase of Beauty

- **Roses (*Rosa spp.*)**: Known for their timeless elegance, roses are popular cut flowers with a diverse range of colors and fragrances.
- **Carnations (*Dianthus caryophyllus*)**: With their ruffled petals and long vase life, carnations are staples in floral arrangements.
- **Tulips (*Tulipa spp.*)**: Tulips' vibrant hues and graceful forms make them beloved choices for various occasions.
- **Lilies (*Lilium spp.*)**: These elegant, fragrant blooms come in various shapes and colors, adding sophistication to arrangements.
- **Alstroemerias (*Alstroemeria spp.*)**: Also known as Peruvian lilies, these long-lasting flowers feature multiple blooms per stem.
- **Sunflowers (*Helianthus annuus*)**: The radiant faces of sunflowers bring warmth and cheer to bouquets with their bold colors and large heads.

The production and packaging of cut flowers is an artful science, involving meticulous cultivation, careful handling, and thoughtful presentation. From nurturing flowers on the farm to adorning homes and events, the journey of cut flowers is a testament to the beauty and fleeting nature of life itself. By understanding the steps involved in producing and caring for cut flowers, we can cherish their beauty and elegance, keeping their vibrant presence alive for longer, enriching our lives and surroundings.

8

Landscape Ornamentals & Shrubs

Landscape ornamentals encompass a wide range of plants that contribute to the aesthetics of a garden or landscape. They include flowering plants, foliage plants, and those with unique shapes or features.

Hydrangeas (Hydrangea spp.):

- ★ Known for their large, colorful flower clusters that change based on soil pH.
- ★ Varieties like 'Endless Summer' offer continuous blooming throughout the season.

Japanese Maple (*Acer palmatum*):

- ★ Renowned for their striking foliage in various shades of red, orange, and purple.
- ★ 'Bloodgood' and 'Sango Kaku' are popular cultivars.

Rose of Sharon (*Hibiscus syriacus*):

- ★ A deciduous shrub with large, showy flowers that attract pollinators.
- ★ Varieties like 'Blue Chiffon' and 'Diana' offer different flower colors.

Crape Myrtle (*Lagerstroemia spp.*):

- ★ Features vibrant, long-lasting flower clusters in shades of pink, red, and white.
- ★ Varieties like 'Natchez' and 'Sioux' are widely planted.

Forsythia (*Forsythia* spp.):

- ★ Early spring bloomer with bright yellow flowers that signal the arrival of spring.
- ★ 'Lynwood Gold' is a popular cultivar.

Shrubs:

Shrubs are woody plants that vary in size and shape, providing structure and often serving as hedges, borders, or standalone features.

1. Boxwood (*Buxus* spp.):

- ★ Evergreen shrubs with small, dense leaves that can be pruned into various shapes.
- ★ 'Green Velvet' and 'Winter Gem' are commonly used cultivars.

2. Lilac (*Syringa* spp.):

- ★ Fragrant spring-blooming shrubs with clusters of purple, pink, or white flowers.
- ★ Varieties like 'Common Lilac' and 'Miss Kim' are well-known.

3. Spirea (*Spiraea* spp.):

- ★ Offers a range of sizes and flower colors, often attracting pollinators.
- ★ 'Goldflame' and 'Bridal Wreath' are popular choices.

4. Viburnum (*Viburnum* spp.):

- ★ Features fragrant flowers, attractive foliage, and sometimes colorful berries.
- ★ 'Snowball' and 'Arrowwood' are commonly planted varieties.

5. Weigela (*Weigela* spp.):

- ★ Spring-blooming shrubs with trumpet-shaped flowers in shades of pink, red, or white.
- ★ 'Wine & Roses' and 'Sonic Bloom' are popular cultivars.

Shade and Ornamental Trees: Adding Elegance to Landscape

Shade and ornamental trees play a pivotal role in landscape design, offering beauty, shade, and structure to outdoor spaces. These trees not only provide relief from the sun's intensity but also contribute to the overall aesthetics and functionality of gardens and parks. In this chapter, we will explore the world of shade and ornamental trees, discussing their characteristics, selection, cultivation, and the transformation they bring to landscapes.

Characteristics of Shade and Ornamental Trees

Shade and ornamental trees bring distinct qualities to landscapes:

- ★ **Canopy Structure**: These trees often have expansive canopies that create shaded areas beneath, providing relief from the sun and defining outdoor spaces.
- ★ **Aesthetic Appeal**: Ornamental trees are chosen for their striking features, including unique foliage, flowers, bark, and growth habits, enhancing the visual appeal of the landscape.
- ★ **Environmental Benefits**: Shade trees offer cooling effects, reduce energy consumption, and improve air quality, making them valuable assets in urban and residential areas.
- ★ **Wildlife Habitat**: Many shade and ornamental trees support wildlife by providing nesting sites, food sources, and shelter.

Selecting and Cultivating Shade and Ornamental Trees

Choosing and caring for shade and ornamental trees involves several considerations:

- ★ **Site Assessment**: Assessing the planting site's soil, sunlight availability, moisture levels, and space constraints is essential for selecting suitable trees.
- ★ **Tree Selection**: Considering factors such as mature size, growth rate, seasonal features (flowers, foliage, and fruit), and adaptability to local climate helps make informed choices.
- ★ **Planting Techniques**: Proper planting practices, including digging an appropriate planting hole, providing proper drainage, and planting at the right depth, set the foundation for healthy growth.
- ★ **Watering**: Young trees require regular and consistent watering during establishment. Once established, deep watering during dry periods ensures their vitality.
- ★ **Mulching**: Applying mulch around the base of the tree conserves soil moisture, regulates temperature, and prevents competition from weeds.

- **Pruning**: Pruning for structural integrity and aesthetics should be done thoughtfully. Removing dead or diseased branches and ensuring proper branch spacing are essential.
- **Tree Care**: Regular monitoring for pests, diseases, and signs of stress, as well as addressing any issues promptly, contributes to the tree's health and longevity.

Noteworthy Shade and Ornamental Trees

- **Japanese Maple (*Acer palmatum*)**: Description: Japanese maples are renowned for their exquisite, finely dissected leaves that come in a variety of colors, including shades of red, orange, green, and even variegated forms. These small to medium-sized trees are ideal for adding a touch of elegance and grace to shaded areas. They are often used as focal points in gardens, and their foliage can transform into fiery shades in the fall.
- **Dogwood (*Cornus spp.*)**: Description: Dogwoods are beloved for their delicate, four-petaled flowers that appear in early spring before the leaves emerge. They come in various species, including the native flowering dogwood (*Cornus florida*) and the kousa dogwood (*Cornus kousa*). The flowers are often white or pink, and some species produce colorful fruit. Dogwoods thrive in partial shade and are perfect for understory planting.
- **Crape Myrtle (*Lagerstroemia spp.*)**: Description: Crape myrtles are known for their stunning, showy flower clusters that bloom in summer, ranging in colors from white and pink to red and purple. These trees also feature attractive peeling bark, which adds visual interest during the winter months. Crape myrtles thrive in full sun and are commonly used in warmer climates to bring vibrant color to landscapes.
- **Redbud (*Cercis spp.*)**: Description: Redbuds are celebrated for their unique, pea-like flowers that appear on bare branches in early spring, before the leaves emerge. The flowers can be pink, purple, or white, depending on the species. Redbuds are often used as ornamental accent trees, and their heart-shaped leaves and attractive branching patterns contribute to their year-round appeal.
- **Weeping Willow (*Salix babylonica*)**: Description: Weeping willows are instantly recognizable for their graceful, cascading branches that sweep down to the ground. These trees thrive in wet or damp areas and are often planted near water features or along riverbanks. They

have slender, lance-shaped leaves that flutter in the breeze, creating a soothing and calming atmosphere.

- **Japanese Flowering Cherry (*Prunus serrulata*)**: Description: Japanese flowering cherries are renowned for their breathtaking spring displays of delicate, fragrant flowers that range in color from white to various shades of pink. These trees are often used as focal points in gardens and are especially popular for creating a stunning floral spectacle during cherry blossom season. Their leaves turn golden-yellow in the fall.

These examples showcase the diversity of shade and ornamental trees that can elevate the beauty and design of landscapes. Each species offers unique attributes, including flowers, foliage, and branching patterns, contributing to the overall aesthetic appeal of outdoor spaces.

The Beauty and Diversity of Foliage Plants: Creating Stunning Green Landscapes

Foliage plants, often overshadowed by their showier flowering counterparts, possess an understated elegance that forms the backbone of many captivating landscapes. With a focus on leaf textures, shapes, colors, and patterns, these plants add depth, contrast, and year-round interest to gardens and interiors. This chapter explores the world of foliage plants, highlighting their unique qualities, cultivation practices, and showcasing a selection of remarkable examples.

The Allure of Foliage Plants

Foliage plants are celebrated for their ability to provide consistent visual interest throughout the year. Their enduring presence, unaffected by the fleeting nature of blooms, makes them essential components of any well-designed garden or indoor space. Key characteristics include:

- **Leaf Diversity**: Foliage plants exhibit a wide range of leaf shapes, sizes, and arrangements. From the large, dramatic leaves of elephant ear plants (Colocasia spp.) to the delicate, feathery fronds of ferns (various genera), the diversity is staggering.
- **Color Palette**: Leaves come in a spectrum of colors, from deep greens to variegated patterns of white, yellow, red, and even shades of purple. This diversity enables gardeners to create striking color combinations.
- **Texture and Form**: The texture of foliage can range from smooth and glossy to coarse and fuzzy. The arrangement of leaves—spiky, rosette-like, or architectural—adds further interest to the overall design.

- ★ **Year-Round Interest**: Unlike flowering plants that may have limited blooming periods, foliage plants offer visual appeal all year long. This ensures a captivating landscape even in the absence of flowers.

Cultivation and Care

Successful cultivation of foliage plants requires careful consideration of their specific needs:

- ★ **Light Requirements**: Foliage plants exhibit varying light preferences, including full sun, partial shade, and full shade. Choosing the appropriate location based on light levels is crucial for maintaining vibrant foliage.
- ★ **Soil Conditions**: Ensure well-draining soil that retains moisture without becoming waterlogged. Amending soil with organic matter can enhance its water-holding capacity and nutrient content.
- ★ **Watering Practices**: Regular, consistent watering is essential for most foliage plants. Avoid overwatering, as this can lead to root rot, and adjust watering frequency based on environmental conditions.
- ★ **Fertilization**: Depending on the plant's growth rate, apply a balanced, slow-release fertilizer during the growing season. Avoid excessive nitrogen, which can lead to lush growth but reduced vibrancy in leaf color.
- ★ **Pruning and Maintenance**: Regularly remove dead or yellowing leaves to maintain the plant's appearance and prevent disease. Pruning can also help control the plant's size and shape.

Remarkable Examples of Foliage Plants

- ★ **Caladium (*Caladium spp.*)**: Known for their vibrant, arrow-shaped leaves, caladiums thrive in shade and come in an array of color combinations.
- ★ **Hostas (*Hosta spp.*)**: Valued for their lush foliage and elegant form, hostas thrive in shaded or partially shaded areas. They offer a range of leaf sizes, textures, and colors.
- ★ **Coleus (*Plectranthus scutellarioides*)**: Renowned for their richly hued leaves and intricate patterns, coleus plants are versatile and can be grown in both containers and garden beds.

- **Japanese Maples (*Acer palmatum*)**: These small trees are prized for their finely dissected leaves that range from vibrant greens to deep purples, adding drama to any landscape.
- **Ferns (*Various Genera*)**: With their graceful, feathery fronds, ferns bring a touch of elegance to shaded areas. They're available in a wide variety of shapes and sizes.
- **Croton (*Codiaeum variegatum*)**: Characterized by their striking, multicolored leaves, crotons are popular indoor plants that thrive in bright, indirect light.

Foliage plants are unsung heroes in the world of gardening, providing a consistent backdrop of beauty that transcends the seasons. By harnessing their diverse leaf forms, colors, and textures, gardeners and indoor plant enthusiasts alike can create breathtaking landscapes that thrive year-round.

9

Artistry in Bloom: Flower Arrangements & Prolonging Vase Life

Flower arrangements are an expression of creativity that transforms blooms into captivating works of art. Beyond aesthetics, mastering the art of arranging flowers involves understanding techniques that enhance the longevity and freshness of the arrangement.

Creating Beauty: The Art of Flower Arrangements

Flower arranging is an art form that has evolved over centuries, reflecting cultural trends, artistic sensibilities, and the natural beauty of the world. In this chapter, we will delve into various techniques used in flower arrangements, showcasing examples that highlight the creative potential of each method.

Ikebana The Elegance of Simplicity

Originating in Japan, Ikebana is a disciplined art form that emphasizes the harmony between nature and humanity. Its minimalist approach focuses on the use of asymmetry, open spaces, and the inherent beauty of each element. A classic example of Ikebana is the "Moribana" style, where different plant materials are arranged in a shallow container. Imagine a Moribana arrangement featuring delicate cherry blossoms and slender bamboo shoots, capturing the essence of tranquility.

English Garden: Effortless Abundance

Inspired by the natural beauty of English gardens, this technique aims to mimic the carefree yet enchanting look of flowers growing wild. Imagine a rustic bouquet containing a medley of roses, peonies, and wildflowers, evoking a sense of a blooming meadow. The key here is to create an arrangement that appears effortlessly gathered, celebrating the beauty of each stem in its natural state.

Formal Linear: Architectural Precision

For those who appreciate order and precision, the formal linear style offers a symmetrical and structured approach. This technique involves arranging flowers and foliage in straight lines, creating a sense of balance and formality. Consider an arrangement of tall calla lilies and long-stemmed gladioli, standing proudly in a tall, slender vase. This style is often favored for formal events and sophisticated settings.

Cascade: Graceful Waterfall of Blooms

The cascade style mimics the flow of a waterfall, with flowers and foliage tumbling gracefully downward. Orchids, amaranthus, and trailing ivy can be combined to create a stunning cascade arrangement that exudes elegance and movement. This technique is often used for weddings and other celebratory occasions, where the arrangement's flowing lines capture attention.

Minimalist Monochrome: Embracing Singular Beauty

In this technique, a single type of flower is used to create an impactful and sophisticated statement. Picture an arrangement of pure white lilies, each stem standing tall in a minimalist vase. This approach highlights the unique characteristics of each bloom while showcasing the power of simplicity.

Hanging Gardens: Elevating Floral Artistry

Taking inspiration from hanging gardens, this technique involves suspending flowers and foliage in midair, creating a dynamic and enchanting display. Delicate blooms like fuchsias, petunias, and trailing vines can be used to craft a stunning hanging arrangement that adds a touch of whimsy to any space.

Abstract Sculptural: Redefining Floral Boundaries

For those seeking avant-garde expressions of floral art, the abstract sculptural technique offers a unique avenue. This method involves using unconventional materials and shapes to create a visually striking arrangement. Imagine an abstract piece incorporating twisted branches, vibrant blooms, and wirework, evoking a sense of mystery and artistic intrigue.

Each of these techniques offers a distinct way of approaching flower arranging, allowing individuals to express their creativity and personal style. Whether you prefer the serene minimalism of Ikebana or the lush exuberance of an English garden, experimenting with these techniques can lead to breathtaking arrangements that captivate the senses and showcase the beauty of nature's bounty.

Flower arrangements are a fusion of design principles, color theory, and personal expression. Key aspects include:

- ★ **Balance and Proportion:** Achieving visual harmony through balanced composition and appropriate scale is essential for creating pleasing arrangements.
- ★ **Color and Contrast:** Thoughtful color selection and contrast between flowers, foliage, and containers contribute to the impact of the arrangement.
- ★ **Texture and Form:** Combining diverse textures and forms, such as spiky, round, and linear elements, adds depth and interest to the arrangement.
- ★ **Focal Points:** Establishing focal points guides the viewer's eye and enhances the arrangement's focal impact.

Prolonging Vase Life: Techniques for Lasting Beauty

While flower arrangements are fleeting, certain practices can extend their beauty:

- ★ **Hydration and Conditioning:** Immediately place cut stems in warm water to hydrate them. Remove leaves below the waterline to prevent bacterial growth.
- ★ **Clean Vases and Tools:** Use clean vases and tools to prevent bacterial contamination. Wash vases with soap and water between arrangements.
- ★ **Recutting Stems:** Re-cut stems at an angle to increase water absorption. Use sharp, clean scissors or pruners to avoid crushing the stems.
- ★ **Water Quality and Flower Food:** Use fresh, clean water and add commercial flower preservatives to nourish blooms and inhibit bacterial growth.
- ★ **Temperature and Environment:** Display arrangements away from direct sunlight, drafts, and heat sources. Cool environments prolong flower freshness.

Inspiring Arrangements: Showcasing Floral Artistry

- **Classic Rose Bouquet:** A timeless arrangement of roses showcases elegance and romance.
- **Wildflower Meadow:** Combining various wildflowers and grasses creates a charming and rustic arrangement.
- **Monochromatic Elegance:** An arrangement composed of flowers in varying shades of a single color exudes sophistication.
- **Tropical Paradise:** Exotic blooms and foliage evoke a sense of the tropics, perfect for adding a bold statement.
- **Modern Minimalism:** A minimalist arrangement with a few carefully chosen blooms and striking design elements embodies contemporary elegance.
- **Garden Gathering:** An arrangement featuring a mix of garden-fresh blooms and foliage reflects the charm of a bountiful garden.

Flower arrangements are a blend of artistic vision and technical expertise, harnessing the fleeting beauty of blooms to create moments of lasting enchantment. By understanding the principles of design, selecting appropriate techniques to extend vase life, and crafting arrangements that resonate with emotions and themes, one can master the art of floral design. Flower arrangements not only bring joy to the present but also celebrate the ephemeral nature of life itself.

10

Integrated Pest and Disease Management for Optimal Flower Crop Harvesting

Effective management of pests and diseases is crucial for the successful cultivation of flower crops. Additionally, proper harvesting techniques are essential to ensure high-quality flowers for commercial or decorative purposes. Here's a general overview of pest and disease management, as well as harvesting practices, for important flower crops, along with some references for further information:

1. Roses (*Rosa* spp.):

- ✯ **Pest Management:** Common pests include aphids, thrips, and spider mites. Integrated pest management (IPM) strategies involving cultural practices, biological control, and targeted pesticide use can be employed.
- ✯ **Disease Management:** Diseases like powdery mildew and black spot can affect roses. Proper spacing, good air circulation, and fungicide application can help manage these issues.
- ✯ **Harvesting:** Roses should be harvested early in the morning when they are well-hydrated. Cut stems at a 45-degree angle and immediately place them in clean water.

2. Chrysanthemums (*Chrysanthemum* spp.):

- ✯ **Pest Management:** Common pests include aphids, thrips, and leafminers. Using insecticidal soaps, neem oil, and beneficial insects can help control these pests.

- ✯ **Disease Management:** Diseases like Botrytis blight and powdery mildew can impact chrysanthemums. Proper sanitation, well-draining soil, and proper spacing can reduce disease incidence.
- ✯ **Harvesting:** Chrysanthemum flowers should be harvested when the outer petals start to open. Cut stems just above a pair of leaves.

3. Carnations (*Dianthus caryophyllus*):

- ✯ **Pest Management:** Aphids, thrips, and mites are common pests for carnations. Biological control and selective pesticide use can be effective.
- ✯ **Disease Management:** Fusarium wilt and bacterial blight can affect carnations. Using disease-resistant varieties and practicing proper sanitation can help manage these diseases.
- ✯ **Harvesting:** Harvest carnation flowers when they are at the "tight bud" stage, just before they start to open. Cut stems in the early morning.

4. Marigold (*Tagetes* spp.):

- ✯ **Pest Management:** Common pests include aphids, whiteflies, and thrips. Implementing cultural practices, using natural predators, and applying appropriate insecticides can help control these pests.
- ✯ **Disease Management:** Marigolds can be susceptible to fungal diseases like powdery mildew and Botrytis. Proper spacing, good air circulation, and fungicide application can help prevent or manage these diseases.
- ✯ **Harvesting:** Marigold flowers can be harvested when fully open. Cut stems in the early morning or late afternoon for the best vase life.

5. Aster (*Aster* spp.):

- ✯ **Pest Management:** Aphids, thrips, and leafhoppers are common pests affecting asters. Encourage beneficial insects and use targeted pesticides to manage pest populations.
- ✯ **Disease Management:** Asters can be vulnerable to diseases like powdery mildew and rust. Proper spacing and good sanitation can help prevent disease spread.
- ✯ **Harvesting:** Harvest aster flowers when they are fully open but before they start to fade. Remove any foliage that will be submerged in water.

6. Lilium (*Lilies*):

- ✯ **Pest Management:** Lily beetles, aphids, and spider mites can affect lilies. Integrated pest management practices and regular scouting can help manage these pests.

- **Disease Management:** Lilies can be prone to diseases such as botrytis and rust. Proper sanitation, well-draining soil, and proper spacing can minimize disease issues.
- **Harvesting:** Harvest lily flowers when they are partially open. Cut stems early in the morning or late in the afternoon for better water uptake.

7. Anthurium (*Anthurium* spp.):

- **Pest Management:** Common pests include mealybugs, scales, and spider mites. Regular monitoring and using appropriate insecticides or natural enemies can help control these pests.
- **Disease Management:** Anthuriums can face bacterial diseases like Xanthomonas leaf spot. Avoid overhead watering and practice good sanitation to prevent disease spread.
- **Harvesting:** Harvest anthurium flowers when the spadix (the central column) is well-formed and the spathe (the colored bract) is fully open. Cut stems carefully to avoid damaging the plant.

Methods of harvesting of major flower crops

Harvesting methods for major flower crops vary based on the specific plant type and the intended use of the flowers. Here are some examples of harvesting methods for different flower crops:

Roses (*Rosa* spp.):

- **Cutting:** Roses are typically cut individually or in small clusters using sharp pruning shears or scissors. Cut stems at a 45-degree angle just above a set of healthy leaves.

Chrysanthemums (*Chrysanthemum* spp.):

- **Disbudding:** For larger blooms, chrysanthemum growers often remove side buds or secondary shoots that develop in the axils of the leaves. This allows the plant to put more energy into producing a single, larger flower per stem.

Carnations (*Dianthus caryophyllus*):

- **Stalk Harvesting:** Carnations are typically harvested by cutting the stems below the calyx (the base of the flower) using sharp shears. Stems should be cut just above a pair of leaves.

Tulips (*Tulipa* spp.):

- ★ **Full Length Cutting:** Tulips are generally harvested when the flower is still in bud and the color is showing. Cut the stem at ground level using a sharp knife or shears.

Sunflowers (*Helianthus* spp.):

- ★ **Full-Length Cutting:** Sunflowers are often harvested when the flower head is fully mature but before the petals start to wilt. Cut the stem at the desired length using sharp shears or a knife.

Lilies (*Lilium* spp.):

- ★ **Partial Opening:** Lilies are typically harvested when the buds are partially open and show color but before they fully open. This helps extend the vase life. Cut stems early in the morning or late in the afternoon.

Gerbera Daisies (*Gerbera* spp.):

- ★ **Stalk Harvesting:** Gerbera daisies are harvested by cutting the stems at the base of the plant. Use sharp scissors or shears to cut the stems cleanly.

Orchids (*Orchidaceae* family):

- ★ **Stalk or Spike Cutting:** For certain orchid types like Phalaenopsis, individual flower spikes are cut when a significant portion of the buds have opened. Cut the spike just above a node to encourage re-blooming.

Anthuriums (*Anthurium* spp.):

- ★ **Stem Cutting:** Anthurium flowers are typically cut with a sharp knife or shears. Cut stems where they meet the main plant stem or at the desired length.

Snapdragons (*Antirrhinum* spp.):

- ★ **Stem Cutting:** Snapdragons are harvested when a few of the flowers on the spike are open and the others are in bud. Cut the stem just above a pair of leaves.

These are general examples of harvesting methods for different flower crops. The specific method and timing might vary based on the intended use, local climate, and the growth stage of the flowers. Always use clean and sharp tools to minimize damage to the plants and ensure longer vase life for the harvested flowers.

Export Potential of the Flower Industry:

The flower industry, also known as the floriculture sector, holds significant export potential due to the global demand for cut flowers, potted plants, and other floral products. Several factors contribute to the export potential of the flower industry:

- **Global Demand:** Flowers and ornamental plants are widely used for various purposes, including decorations, gifts, events, and religious ceremonies, driving consistent global demand.
- **Special Occasions:** Flowers are associated with special occasions such as weddings, birthdays, anniversaries, and holidays, leading to seasonal peaks in demand.
- **Cultural Significance:** Flowers hold cultural and symbolic importance in many societies, leading to consistent demand in different parts of the world.
- **Increasing Urbanization:** As urbanization increases, there is a growing demand for indoor plants and ornamental gardening, boosting the export potential of potted plants and indoor flowers.
- **Growing Middle Class:** Rising incomes and an expanding middle class in emerging economies contribute to increased spending on non-essential items like flowers.
- **Innovation:** The floriculture industry has embraced innovation, such as new flower varieties, sustainable growing practices, and advanced post-harvest technologies, making exports more attractive.

Challenges of the Flower Industry:

While the flower industry has great potential for exports, it also faces several challenges that impact its growth and profitability:

- **High Production Costs:** Flowers require specific growing conditions and care, leading to high production costs for labor, energy, transportation, and technology.
- **Seasonality:** Many flowers have specific growing seasons, leading to challenges in meeting year-round demand and maintaining consistent export volumes.
- **Perishability:** Flowers are highly perishable, requiring careful handling, refrigeration, and transportation, which can increase logistics costs.

- ✯ **Competition:** The flower industry is highly competitive, with major flower-producing countries vying for market share in the global export market.
- ✯ **Market Access and Regulations:** Exporting flowers often involves navigating complex import regulations, phytosanitary requirements, and trade agreements in various countries.
- ✯ **Environmental Concerns:** The environmental impact of flower production, such as water usage and pesticide application, raises concerns in terms of sustainability and ethical practices.
- ✯ **Price Fluctuations:** The flower market can experience price fluctuations due to factors like changes in supply, demand, and currency exchange rates.
- ✯ **Transportation Challenges:** The need for timely transportation and access to direct flights can be challenging, especially for exporting perishable flowers to distant markets.
- ✯ **Consumer Preferences:** Shifts in consumer preferences, trends, and demand for specific flower varieties can impact the market demand for certain products.

Despite these challenges, the flower industry continues to thrive due to the emotional and aesthetic value flowers hold for people across the globe. Addressing these challenges through innovation, sustainable practices, and strategic market development can help flower exporters tap into the industry's export potential.

"In a Nutshell: Rapid Q&A"

1. What is the process of artificially inducing flowering and fruiting in plants called?

Ans. Fruit Induction or Flower Induction.

2. Which plant hormone is responsible for fruit ripening?

Ans. Ethylene.

3. What is the scientific term for seedless fruits?

Ans. Parthenocarpic Fruits.

4. What type of fruit develops from a single flower with multiple carpels?

Ans. Aggregate Fruit.

5. Which gas is used to delay fruit ripening during storage?

Ans. Carbon Dioxide (CO_2).

6. Which fruit is known as the "king of fruits"?

Ans. Mango.

7. What is the edible part of a strawberry commonly called?

Ans. Receptacle.

8. Which fruit is botanically classified as a berry?

Ans. Tomato.

9. What type of fruit has a hard, woody shell and is indehiscent?

Ans. Nut.

10. The outermost layer of a citrus fruit's peel is called what?

Ans. Flavedo or Zest.

11. What is the term for the process of preventing fruit from forming and growing on a tree?

Ans. Fruit Thinning.

12. Which fruit is rich in vitamin C and is often used to make juice?

Ans. Orange.

13. The scientific name for the apple genus is what?

Ans. Malus.

14. What is the primary function of the mesocarp in a fruit?

Ans. Fleshy tissue that surrounds the seed.

15. What type of fruit develops from multiple ovaries of separate flowers in a single inflorescence?

Ans. Multiple Fruit.

16. What is the term for the practice of grafting two different fruit tree varieties onto the same rootstock?

Ans. Double Grafting or Double Working.

17. What is the hormone responsible for apical dominance in plants?

Ans. Auxin.

18. Which gas is released by fruits during respiration and promotes ripening?

Ans. Ethylene.

19. What is the process of removing water from fruits through controlled drying called?

Ans. Dehydration.

20. What is the term for the hybrid fruit resulting from the cross between a tangerine and a pomelo?

Ans. Tangelo.

21. The part of a banana bunch that protects the emerging bananas is called what?

Ans. Bract or Spatha.

22. Which fruit is commonly called the "stone fruit" due to its hard pit?

Ans. Peach.

23. What is the process of transferring pollen from one flower to the stigma of another flower called?

Ans. Pollination.

24. The process of applying a solution to break seed dormancy is known as what?

Ans. Scarification.

25. What is the term for a fruit that splits open to release its seeds when mature?

Ans. Dehiscent Fruit.

26. Which fruit is a type of berry with a thick rind and numerous seeds?

Ans. Watermelon.

27. What is the term for the edible part of a pineapple that is made up of fused berries?

Ans. Multiple Fruit or Syncarp.

28. What is the phenomenon where a plant produces fruit without fertilization called?

Ans. Parthenocarpy.

29. What is the process of artificially breaking seed dormancy by exposing them to cold temperatures called?

Ans. Cold Stratification.

30. Which fruit is botanically classified as a pome?

Ans. Apple.

31. What is the specialized underground stem of a potato plant called?

Ans. Tuber.

32. What is the term for a fruit that does not split open to release its seeds?

Ans. Indehiscent Fruit.

33. Which fruit is often called the "banana plantain"?

Ans. Plantain.

34. What is the scientific name for the grape genus?

Ans. Vitis.

35. What is the part of the flower that eventually becomes the fruit called?

Ans. Ovary.

36. Which fruit is known for its distinctive smell and spiky outer shell?

Ans. Durian.

37. What is the term for the process of grafting a scion onto a rootstock of the same plant species?

Ans. Homografting.

38. Which fruit is known as the "queen of fruits"?

Ans. Mangosteen.

39. What type of fruit develops from a single ovary of a single flower?

Ans. Simple Fruit.

40. What is the term for the process of removing excess fruit from a tree to improve quality and yield?

Ans. Fruit Thinning.

41. Which fruit is commonly known as the "custard apple"?

Ans. Annona.

42. What is the edible part of a pineapple known as?

Ans. Flesh or Pineapple Pulp.

43. What is the process of protecting developing fruits from damage by placing bags or covers over them called?

Ans. Bagging.

44. Which fruit is botanically classified as a drupe?

Ans. Cherry.

45. What is the scientific name of the apple?

Ans. *Malus domestica.*

46. Which gas is responsible for the ripening of fruits?

Ans. Ethylene.

47. What is the process of artificially controlling the dormancy of seeds called?

Ans. Stratification.

48. What is the term for the process of transferring pollen from the anther to the stigma?

Ans. Pollination.

49. Which vitamin is abundant in citrus fruits?

Ans. Vitamin C.

50. What is the process of grafting a shoot onto a rootstock called?

Ans. Budding.

51. Which fruit is known as the "king of fruits"?

Ans. Mango.

52. What is the hormone responsible for the elongation of plant cells, promoting growth?

Ans. Gibberellin.

53. Which fruit has the highest water content?

Ans. Watermelon.

54. What is the term for the process of removing excess flowers or fruits to enhance quality and size?

Ans. Thinning.

55. Which fruit is considered a berry: banana or peach?

Ans. Banana.

56. What is the edible part of a strawberry known as?

Ans. Accessory fruit or aggregate fruit.

57. Which gas is essential for photosynthesis?

Ans. Carbon dioxide (CO_2).

58. Which fruit is a good source of dietary fiber and is often recommended for digestion?

Ans. Papaya.

59. Which hormone is responsible for the abscission (shedding) of leaves and fruits?

Ans. Ethylene.

60. What is the scientific name of the banana?

Ans. Musa spp.

61. Which fruit is known as the "queen of fruits"?

Ans. Pineapple.

62. Which nutrient is essential for fruit color development and is often associated with antioxidants?

Ans. Vitamin A.

63. Which fruit is commonly associated with Halloween, often carved as decorations?

Ans. Pumpkin.

64. What is the process of grafting a rootstock onto a shoot called?

Ans. Inarching.

65. Which fruit has a tough outer rind and is known for its juicy, segmented interior?

Ans. Citrus fruits (e.g., oranges, grapefruits).

66. What is the process of breaking seed dormancy by subjecting seeds to low temperatures called?

Ans. Cold stratification.

67. Which fruit is rich in potassium and known for its ability to lower blood pressure?

Ans. Banana.

68. What is the term for the study and cultivation of fruits?

Ans. Pomology.

69. Which hormone promotes fruit ripening and softening?

Ans. Ethylene.

70. Which fruit is known as the "stone fruit" and includes peaches, plums, and cherries?

Ans. Drupes.

71. What is the process of removing the terminal bud to encourage lateral growth called?

Ans. Pinching.

72. Which fruit is considered a member of the rose family?

Ans. Apple.

73. What is the layer of cells responsible for secondary growth in stems and roots?

Ans. Cambium.

74. Which fruit has a rough exterior and sweet, aromatic flesh and is often used in desserts?

Ans. Pineapple.

75. What is the process of artificially controlling the light exposure of plants to induce flowering called?

Ans. Photoperiod manipulation.

76. Which hormone is responsible for the closure of stomata during water stress?

Ans. Abscisic acid.

77. What is the term for the process of transferring pollen from one flower to another flower on a different plant?

Ans. Cross-pollination.

78. Which fruit is known as the "crowning glory" of fruits due to its distinctive appearance?

Ans. Strawberry.

79. What is the term for the loss of water vapor from plant surfaces?

Ans. Transpiration.

80. Which fruit is rich in antioxidants, particularly lycopene?

Ans. Tomato.

81. What is the term for the branch less trunk of a tree that produces fruit?

Ans. Stem or trunk.

82. Which fruit is often used as a laxative due to its high fiber content?

Ans. Fig.

83. What is the pigment responsible for the green color of plants?

Ans. Chlorophyll.

84. Which fruit is known for its soft, creamy texture and is often used in desserts and smoothies?

Ans. Avocado.

85. What is the process of removing flower buds to redirect energy towards vegetative growth called?

Ans. De-blossoming.

86. Which hormone is responsible for the promotion of fruit setting and prevents fruit drop?

Ans. Auxin.

87. What is the scientific name of the citrus fruit family?

Ans. Rutaceae.

88. Which fruit is considered a natural diuretic and is known for its cleansing properties?

Ans. Watermelon.

89. What is the process of promoting flowering by exposing plants to low temperatures called?

Ans. Vernalization.

90. Which fruit is commonly used to make jams, jellies, and beverages?

Ans. Grapes.

91. What is the hormone responsible for promoting cell elongation and growth in young plants?

Ans. Auxin.

92. Which fruit is known for its distinctive star-like shape and sweet-tart taste?

Ans. Starfruit (carambola).

93. What is the term for the protective outer covering of a seed?

Ans. Seed coat.

94. Which fruit is known as the "fruit of paradise" and is associated with a tropical island?

Ans. Mango.

95. What is the process of grafting used for in vegetable science?

Ans. Grafting is used to combine the desirable traits of one plant onto the rootstock of another.

96. What term describes the cultivation of vegetables in water without soil?

Ans. Hydroponics is the method of cultivating vegetables in a soil less environment using nutrient-rich water.

97. Which gas is released by plants during photosynthesis?

Ans. Oxygen is released by plants as a byproduct of photosynthesis.

98. What is the study of vegetable genetics known as?

Ans. Vegetable genetics is known as horticultural genetics.

99. What are the essential nutrients required for healthy vegetable growth?

Ans. Essential nutrients include nitrogen, phosphorus, and potassium (NPK), as well as micro nutrients like iron and zinc.

100. What vegetable is known as "*Brassica oleracea*" and has various cultivars like broccoli and cabbage?

Ans. Kale is a vegetable that belongs to the species *Brassica oleracea*.

101. Which plant hormone is responsible for promoting cell elongation in vegetable growth?

Ans. Auxins are plant hormones responsible for promoting cell elongation and growth.

102. What type of vegetables belong to the Solanaceae family?

Ans. Tomatoes, peppers, and eggplants belong to the Solanaceae family.

103. What is the term for the process of controlled exposure to cold temperatures to stimulate seed germination?

Ans. Stratification is the process of exposing seeds to cold temperatures to break dormancy and encourage germination.

104. Which vegetable is classified as a root crop and is a good source of carbohydrates?

Ans. Carrots are root crops that are rich in carbohydrates.

105. What is the technique of pinching used for in vegetable cultivation?

Ans. Pinching involves removing the growing tip of a plant to encourage branching and bushier growth.

106. Which gas is essential for photosynthesis to occur in vegetables?

Ans. Carbon dioxide (CO_2) is necessary for photosynthesis to take place in vegetables.

107. What is the primary purpose of using row covers in vegetable farming?

Ans. Row covers protect vegetables from pests, frost, and extreme weather conditions.

108. What is the concept of monocotyledonous and dicotyledonous vegetables based on?

Ans. It refers to the number of seed leaves (cotyledons) present in the embryo of a germinating seed.

109. Which vegetable is known as *"Allium sativum"* and is known for its pungent flavor?

Ans. Garlic is a vegetable known as *"Allium sativum."*

110. What is the practice of rotating crops to prevent soilborne diseases called?

Ans. Crop rotation is the practice of changing the type of crops grown in a particular area to prevent diseases and maintain soil fertility.

111. Which mineral nutrient deficiency is characterized by yellowing of leaves in vegetables?

Ans. Iron deficiency leads to yellowing of leaves, a condition known as chlorosis.

112. What is the term for the process of transferring pollen from the male to the female part of a flower within the same plant?

Ans. Self-pollination is the process of transferring pollen within the same flower or plant.

113. What vegetable is scientifically known as *"Cucurbita pepo"* and includes varieties like zucchini and pumpkin?

A. Squash is a vegetable classified as *"Cucurbita pepo."*

114. Which plant hormone is responsible for promoting fruit ripening in vegetables?

Ans. Ethylene is a plant hormone responsible for promoting fruit ripening.

115. What is the term for the practice of trimming leaves to direct a plant's energy towards fruit development?

Ans. Pruning is the practice of trimming leaves to promote fruit development and improve air circulation.

116. What is the process of hardening off in vegetable cultivation?

Ans. Hardening off involves gradually acclimating seedlings to outdoor conditions to prevent transplant shock.

117. Which vegetable is classified as a member of the cruciferous family and is rich in vitamin C?

Ans. Broccoli is a cruciferous vegetable that is high in vitamin C.

118. What is the scientific name for the common tomato?

Ans. The scientific name for tomato is "*Solanum lycopersicum.*"

119. Which vegetable is known as "Beta vulgaris" and has varieties like sugar beet and chard?

Ans. Beetroot is a vegetable known as "Beta vulgaris."

120. What is the term for the protective layer that covers a seed?

Ans. The seed coat is the protective layer that covers a seed.

121. What vegetable is known as "*Lactuca sativa*" and is often used in salads?

Ans. Lettuce is a vegetable known as "*Lactuca sativa.*"

122. What is the practice of thinning in vegetable cultivation?

Ans. Thinning involves removing excess seedlings to provide adequate space for healthy growth.

123. Which plant nutrient deficiency leads to poor fruit development and blossom end rot?

Ans. Calcium deficiency can result in poor fruit development and blossom end rot in vegetables.

124. What is the term for the process of removing the male reproductive organs of a flower to prevent self-pollination?

Ans. Emasculation is the process of removing the stamens from a flower to prevent self-pollination.

125. Which vegetable is known as "*Cucumis sativus*" and belongs to the gourd family?

Ans. Cucumber is a vegetable known as "*Cucumis sativus.*"

126. What is the term for the first stage of seed growth, where the embryonic root emerges?

Ans. Germination is the process where the embryonic root (radicle) emerges from the seed.

127. Which vegetable is commonly referred to as an "aubergine" and is botanically known as "*Solanum melongena*"?

Ans. Eggplant is a vegetable known as "*Solanum melongena.*"

128. What is the process of acclimating plants to a new environment before transplanting called?

Ans. Hardening off is the process of acclimating plants to outdoor conditions before transplanting.

129. Which vegetable is known as "*Capsicum annuum*" and comes in various colors and heat levels?

Ans. Bell pepper is a vegetable known as "*Capsicum annuum.*"

130. What is the term for the practice of training plants to grow in a specific shape or form?

Ans. Topiary is the practice of training plants into specific shapes or forms.

131. Which plant hormone is responsible for promoting root growth in vegetables?

Ans. Gibberellins are plant hormones responsible for promoting root growth.

132. What is the term for the outermost layer of a stem that protects the underlying tissues?

Ans. The epidermis is the outermost layer of a stem that provides protection.

133. Which vegetable is scientifically known as "*Raphanus sativus*" and is known for its pungent root?

Ans. Radish is a vegetable known as "*Raphanus sativus.*"

134. What is the practice of removing lateral shoots or branches called?

Ans. The practice of removing lateral shoots is called "suckering."

135. Which plant nutrient is responsible for promoting green leafy growth in vegetables?

Ans. Nitrogen is a nutrient that promotes green leafy growth in vegetables.

136. What is the term for the process of transferring pollen from the male to the female part of a flower between different plants?

Ans. Cross-pollination is the process of transferring pollen between different plants.

137. Which vegetable is known as "*Zea mays*" and is a cereal crop with edible kernels?

Ans. Sweet corn is a vegetable known as "*Zea mays.*"

138. What is the term for the swollen underground stem of a plant that stores energy?

Ans. A tuber is a swollen underground stem that stores energy and nutrients.

139. Which vegetable is classified as *"Allium cepa"* and is widely used in cooking?

Ans. Onion is a vegetable classified as *"Allium cepa."*

140. What is the term for the process of transferring a seedling from a small container to a larger one before transplanting?

Ans. Potting up is the process of moving a seedling to a larger container before transplanting.

141. Which vegetable is known as *"Cucumis melo"* and includes varieties like cantaloupe and honeydew?

Ans. Melon is a vegetable known as *"Cucumis melo."*

142. What is the term for the protective layer that encloses the embryo of a seed?

Ans. The endosperm is the protective layer that encloses the embryo of a seed.

143. Which vegetable is commonly referred to as a "green bean" and is scientifically known as *"Phaseolus vulgaris"*?

Ans. Green bean is a vegetable known as *"Phaseolus vulgaris."*

144. What is the cultivation of flowers and ornamental plants called?

Ans. Floriculture is the cultivation of flowers and ornamental plants.

145. Which flower is known as the "Queen of Flowers"?

Ans. The rose is often referred to as the "Queen of Flowers."

146. What term describes the art of designing and arranging flowers aesthetically?

Ans. Flower arranging is known as floral design.

147. Which gas is released by plants during photosynthesis and is essential for flower growth?

Ans. Oxygen is released during photosynthesis and is vital for flower growth.

148. What is the term for the process of inducing dormancy in flower bulbs?

Ans. Vernalization is the process of inducing dormancy in flower bulbs through cold treatment.

149. What is the practice of removing spent flowers to encourage more blooms called?

Ans. Deadheading is the practice of removing faded flowers to promote further flowering.

150. Which flower is associated with the "Language of Flowers" due to its varied symbolic meanings?

Ans. The iris is often linked to the "Language of Flowers" with multiple symbolic meanings.

151. What type of flowers open and close in response to light changes?

Ans. Phototropic flowers open and close in response to changes in light.

152. What is the term for the small flowers that make up the central part of a composite flower head?

Ans. Florets are the small flowers that compose the central part of a composite flower head.

153. Which hormone is responsible for inhibiting lateral bud growth in plants?

Ans. Apical dominance is regulated by auxins, inhibiting lateral bud growth.

154. What is the term for the process of preserving flowers by removing moisture with a drying agent?

Ans. Desiccation is the process of preserving flowers by using a drying agent.

155. Which flower is known for its vibrant colors and trumpet-like shape?

Ans. The petunia is known for its trumpet-shaped flowers and vibrant colors.

156. What is the term for a flower with both male and female reproductive organs?

Ans. A perfect flower has both male (stamen) and female (pistil) reproductive organs.

157. Which flower is associated with November and often symbolizes remembrance?

Ans. The chrysanthemum is associated with November and symbolizes remembrance.

158. What is the term for the specialized underground stem that stores nutrients for future growth?

Ans. A bulb is a specialized underground stem that stores nutrients and energy.

159. Which flower is often used in religious ceremonies and symbolizes purity?

Ans. The white lily is frequently used in religious ceremonies and represents purity.

160. What is the practice of cutting young flower shoots to promote branching and bushier growth?

Ans. Pinching is the practice of cutting young shoots to encourage branching.

161. Which flower is known as "Dianthus caryophyllus" and is prized for its fragrant blooms?

Ans. The carnation is known as "Dianthus caryophyllus" and has fragrant flowers.

162. What is the term for the study of flowers and flowering plants?

Ans. The study of flowers is called floristics or angiosperm taxonomy.

163. Which hormone is responsible for flower bud initiation and development?

Ans. Cytokinins play a role in initiating and developing flower buds.

164. What is the term for the process of crossbreeding two different varieties of flowers?

Ans. Hybridization is the process of crossbreeding different flower varieties.

165. Which flower is known as "*Helianthus annuus*" and is famous for its large, yellow blooms?

Ans. The sunflower is known as "*Helianthus annuus*" and has large, yellow flowers.

166. What is the term for the outermost whorl of floral organs that protects the inner parts?

Ans. The calyx is the outermost whorl of floral organs, often green and protective.

167. Which flower is often used in bouquets and is available in a wide range of colors?

Ans. The gerbera daisy is commonly used in bouquets and comes in various colors.

168. What is the term for the structure that connects the stigma and ovary of a flower?

Ans. The style is the structure that connects the stigma and ovary of a flower.

169. Which flower is associated with love and has varieties like "Grand Prix" and "Freedom"?

Ans. The red rose is associated with love and is known for its varieties.

170. What is the practice of grafting two different flower plants together called?

Ans. Budding is the practice of grafting two different plants, usually related species.

171. Which flower is often used to represent beauty, grace, and admiration?

Ans. The orchid is often associated with beauty, grace, and admiration.

172. What is the term for a flower with only male or female reproductive organs?

Ans. Imperfect flowers have either male or female reproductive organs, not both.

173. Which hormone is responsible for promoting flower opening and fruit development?

Ans. Ethylene is a hormone that promotes flower opening and fruit ripening.

174. What is the term for a flower cluster with a main stalk and multiple individual flowers?

Ans. An inflorescence is a flower cluster composed of multiple individual flowers.

175. Which flower is known as "Narcissus" and is often associated with spring?

Ans. The daffodil is known as "Narcissus" and is a spring-flowering bulb.

176. What is the term for the female reproductive part of a flower that contains ovules?

Ans. The pistil is the female reproductive part containing ovules and stigma.

177. Which flower is associated with purity and is a common choice for bridal bouquets?

Ans. The white lily is often chosen for bridal bouquets and symbolizes purity.

178. What is the term for the process of inducing flowering by exposing plants to cold temperatures?

Ans. Vernalization is the process of promoting flowering through cold treatment.

179. Which flower is known as "*Ranunculus*" and is recognized for its multilayered petals?

Ans. The buttercup is known as "*Ranunculus*" and has multilayered petals.

180. What is the term for the male reproductive organ of a flower that produces pollen?

Ans. The stamen is the male reproductive organ that produces pollen.

181. Which flower is often associated with tranquility and is found in water gardens?

Ans. The water lily is associated with tranquility and thrives in water gardens.

182. What is the practice of preventing flowering to redirect energy towards vegetative growth called?

Ans. Disbudding is the practice of removing flower buds to promote vegetative growth.

183. Which flower is known as "Lilium" and comes in various colors and patterns?

Ans. The lily is known as "Lilium" and has a wide range of colors and patterns.

184. What is the term for the female reproductive organ of a flower that receives pollen?

Ans. The stigma is the female reproductive organ that receives pollen.

185. Which flower is often associated with eternal love and is used in many wedding ceremonies?

Ans. The white orchid is associated with eternal love and is popular in weddings.

186. What is the practice of manipulating light and temperature to control flowering in plants called?

Ans. Photoperiodism is the manipulation of light and temperature to regulate flowering.

187. Which flower is known as "Tulipa" and is famous for its vibrant, cup-shaped blooms?

Ans. The tulip is known as "Tulipa" and has distinctive cup-shaped flowers.

188. What is the term for the part of a flower that supports the petals and reproductive organs?

Ans. The receptacle is the part of a flower that supports the petals and reproductive organs.

189. Which flower is associated with happiness and is often gifted on various occasions?

Ans. The daisy is associated with happiness and is commonly gifted.

190. What is the term for a flower that has both male and female reproductive parts but on separate plants?

Ans. Dioecious plants have separate male and female flowers on different plants.

191. Which flower is often used for its fragrant oil and is associated with relaxation?

Ans. Lavender is used for its fragrant oil and is associated with relaxation.

192. What is the term for the process of artificially inducing flower bud formation out of season?

Ans. Forcing is the process of inducing flower bud formation outside the natural season.

193. Which flower is known as "Camellia" and is admired for its glossy evergreen leaves and blooms?

Ans. The camellia is known as "Camellia" and is appreciated for its leaves and flowers.

194. What's the purpose of pinching in floriculture?

Ans. To promote branching and produce more flowers.

195. Which flowering plants are commonly used for commercial cut flower production?

Ans. Roses, carnations, chrysanthemums, and lilies.

196. What's hydroponic floriculture?

Ans. Growing flowers without soil, using nutrient-rich water solutions.

197. How does temperature influence flower development in floriculture?

Ans. Temperature affects flowering time, flower size, and color.

198. What's the role of ethylene gas in floriculture?

Ans. It can cause premature wilting and aging of flowers.

199. What's the economic importance of floriculture?

Ans. It contributes to the multi-billion dollar global floral industry for events, decorations, and gifts.

Glossary

Pomology: The science and practice of growing and cultivating fruit crops.

Grafting: The process of joining a scion (desired plant) onto a rootstock (base plant) to create a new plant.

Pollination: The transfer of pollen from the male part of the flower to the female part, leading to fruit development.

Thinning: The removal of excess fruit from a tree to improve fruit quality and prevent overcrowding.

Dormancy: A period of reduced physiological activity in fruit trees during cold weather.

Pruning: The controlled removal of plant parts (branches, shoots) to shape and manage growth.

Fruit Set: The process where fertilized flowers develop into small fruit structures.

Cross-Pollination: The transfer of pollen between flowers of different plants, promoting genetic diversity.

Self-Pollination: Pollination that occurs within a single flower or plant.

Pest Management: Strategies and techniques used to control and prevent pest infestations in fruit crops.

Olericulture: The branch of horticulture focused on the cultivation of vegetable crops.

Cultivar: A cultivated variety of a plant species developed for specific traits or characteristics.

Companion Planting: Planting different crops together to maximize growth or deter pests.

Mulching: Applying a protective layer of organic or inorganic material to the soil surface to conserve moisture and control weeds.

Crop Rotation: The practice of changing the type of crops grown in a particular area to prevent soilborne diseases and maintain soil fertility.

Transplanting: Moving seedlings from a nursery to their final growing location.

Row Spacing: The distance between rows of vegetable crops in a field.

Staking: Using supports to keep vegetable plants upright and off the ground.

Hybrid: A plant resulting from the crossbreeding of two different varieties or species.

Cover Crop: A crop grown primarily to benefit the soil and ecosystem rather than for harvest.

Floriculture: The branch of horticulture focused on the cultivation of flowers and ornamental plants.

Bouquet: An arrangement of cut flowers or ornamental plants.

Bulb: A specialized underground storage structure that contains nutrients and energy for plant growth.

Cutting: A plant propagation method involving the removal of a piece of a plant that can develop roots and grow into a new plant.

Petuni A: A popular flowering plant known for its colorful trumpet-shaped blooms.

Topiary: The art of shaping and trimming plants into ornamental forms.

Vase Life: The period of time cut flowers remain attractive in a vase or container.

Greenhouse: A controlled environment structure used for growing flowers and plants under controlled conditions.

Nursery: A facility where plants are grown, nurtured, and sold, often as seedlings or young plants.

Perennial: A plant that lives for more than two years and often produces flowers.

Post-Harvest Technology: The techniques and processes used to handle, store, and preserve harvested crops.

Cool Chain: The uninterrupted temperature-controlled process that extends from harvesting to distribution to maintain product quality.

Dehydration: The removal of moisture from fruits and vegetables to extend shelf life.

Ethylene: A natural plant hormone that influences ripening and can be used for artificial ripening of fruit.

Cold Storage: Refrigerated storage facilities used to prolong the shelf life of fruits and vegetables.

Modified Atmosphere Packaging (MAP): Packaging technique that alters the composition of gases around produce to extend freshness.

Quality Control: Measures taken to ensure that harvested produce meets established quality standards.

Washing: Cleaning harvested crops to remove dirt and contaminants before distribution.

Waxing: Applying a thin layer of wax to fruits and vegetables to enhance appearance and preservation.

Shelf Life: The duration a product remains acceptable for consumption or sale after harvest.

Gardening is how I relax. It's another form of creating and playing with colors."

— Oscar de la Renta, Dominican Fashion Designer

Index